ADVANCES IN
NUMERICAL HEAT TRANSFER

Volume 5

Series in Computational and Physical Processes in Mechanics and Thermal Sciences

Series Editors

W. J. Minkowycz

Mechanical and Industrial Engineering
University of Illinois at Chicago
Chicago, Illinois

E. M. Sparrow

Mechanical Engineering
University of Minnesota
Minneapolis, Minnesota

ADVANCES IN NUMERICAL HEAT TRANSFER

Volume 5

NUMERICAL SIMULATION OF HEAT EXCHANGERS

Edited by

W. J. Minkowycz
Mechanical and Industrial Engineering
University of Illinois at Chicago
Chicago, Illinois

E. M. Sparrow
Mechanical Engineering
University of Minnesota
Minneapolis, Minnesota

J. P. Abraham
School of Engineering
University of St. Thomas
St. Paul, Minnesota

J. M. Gorman
Mechanical Engineering
University of Minnesota
Minneapolis, Minnesota

CRC Press
Taylor & Francis Group
Boca Raton London New York

CRC Press is an imprint of the
Taylor & Francis Group, an **informa** business

CRC Press
Taylor & Francis Group
6000 Broken Sound Parkway NW, Suite 300
Boca Raton, FL 33487-2742

First issued in paperback 2019

© 2017 by Taylor & Francis Group, LLC
CRC Press is an imprint of Taylor & Francis Group, an Informa business

No claim to original U.S. Government works

ISBN-13: 978-0-4822-5019-0 (hbk)
ISBN-13: 978-0-367-87037-9 (pbk)

Visit the Taylor & Francis Web site at
http://www.taylorandfrancis.com

and the CRC Press Web site at
http://www.crcpress.com

Contents

Preface

Heat exchanger design is closely linked to the nature of the participating fluid flows. In turn, the nature of the flow is strongly affected by the characteristics of the fluid mover that delivers the flow. There is substantial evidence that the pressure–flow characteristics of a fluid mover are affected by the flow resistance of the heat exchanger, so that the manufacturer-supplied fan or blower curves are, at best, approximate. This state of affairs serves to motivate the work reported in Chapter 1, which advocates treating the fluid mover and the heat exchanger as a single system. The chapter applies this approach to several real-world heat exchanger situations and demonstrates the errors that occur when the system concept is not utilized.

In Chapter 2, various computational approaches for the analysis of transport phenomena are briefly summarized and recent works are cited. Use of computational fluid dynamics (CFD) methods for complex fluid flow and heat transfer situations are described for important engineering applications, including plate heat exchangers, radiators, and heat recovery units. The results reveal that the CFD approach can reveal important physical processes as well as provide satisfactory results when compared with corresponding experiments. Neural network configurations are introduced along with a case study illustrating the validation of the network based on experimentally measured databases for Nusselt number and friction factor for three kinds of heat exchangers. A generic algorithm for thermal design and optimization of a compact heat exchanger is described. This generic algorithm can provide good auto-search and optimization capabilities in the thermal design of heat exchangers without the use of a trial-and-error process.

Chapter 3 primarily focuses on numerical studies performed to analyze a plate heat exchanger composed of specific types of plates and the experimental validation of the numerical predictions. Different turbulent models, mainly k-ε, RNG k-ε, EARSM k-ε, k-ω, and SST k-ω, are used for the simulations and validations. The effects of several geometrical properties on exchanger performance are investigated numerically with the help of the validated models. The geometrical parameters include channel height, wave amplitude, and distribution channel layout. New plate designs with improved thermal and hydraulic performance are obtained.

Chapter 4 presents a state-of-the-art overview of numerical methods for single-phase and two-phase flows in microchannel heat exchangers. Micro heat exchangers are characterized by at least one of the participating fluids being confined in channels or tubes with typical dimensions of less than 1 mm. Governing equations are conveyed for both these types of flow, with and without phase change. Characteristics of multiphase modeling approaches (i.e., Eulerian–Eulerian, Eulerian–Lagrangian, and direct numerical simulation) are compared. The advantages and disadvantages of several continuum methods for interface evolution (e.g., volume of fluid, level set, phase field, front tracking, and moving mesh) are discussed along with the mesoscopic lattice Boltzmann method. Methods to deal with the mass nonconservation in the level set method are provided.

In Chapter 5, an overview of the analysis and numerical simulations of various types of heat pipes under a variety of operating conditions is presented. The significant and rapid progression of heat pipe technology is examined from the perspective of state-of-the-art modeling and the full simulation that has become possible in the last few decades. Simulations can accurately predict the thermal performance of heat pipes under various operating conditions, including steady-state, continuum transient, and frozen start-up solutions, despite the associated complex multiphase and multidomain transport phenomena. Steady-state and transient heat pipe simulations involve conjugate heat transfer involving the wall, wick, and vapor, although pulsating and loop heat pipes require more fundamental research efforts to explain the physical phenomena of these devices. Furthermore, simulation of the liquid–vapor interface requires consideration of the multiphase phenomena within various wick structures and will provide future understanding and prediction of the heat transport limitation in heat pipes.

Editors

W. J. Minkowycz is the James P. Hartnett professor of mechanical engineering at the University of Illinois at Chicago. He joined the faculty at UIC in 1966. His primary research interests lie in the numerical modeling of fluid flow and heat transfer problems. He has performed seminal research in several branches of heat transfer and has published more than 175 papers in archival journals, in addition to winning numerous awards for his excellence in teaching, research, and service to the heat transfer community. Professor Minkowycz is also editor-in-chief of the *International Journal of Heat and Mass Transfer, Numerical Heat Transfer*, and *International Communications in Heat and Mass Transfer*.

E. M. Sparrow is a member of the National Academy of Engineering and a professor of mechanical engineering at the University of Minnesota, Minneapolis, Minnesota. He has worked on heat transfer problems for more than 60 years, starting at the Raytheon Company and continuing at the National Advisory Committee for Aeronautics Lewis Flight Propulsion Laboratory before coming to the University of Minnesota. He has published more than 700 peer-reviewed articles in the field of engineering and has directed more than 100 doctoral theses to completion. His current interests include industrial applications and real-world problems in addition to numerical heat transfer and fluid flow. He is a member of the National Academy of Engineering.

J. P. Abraham is a professor of mechanical engineering at the University of St. Thomas in St. Paul, Minnesota. His research activities extend broadly over several facets of heat transfer, climatology, fluid flow, biomedical engineering, and renewable energy. In addition to his work in heat transfer, he is also actively engaged with climatology and biomedical engineering. Currently, he has about 230 journal publications, and numerous conference presentations, book chapters, and patents. He is an established spokesman for scientists and engineers who are concerned about the long-term effects of climate change on sustainability.

J. M. Gorman is a research associate at the University of Minnesota, Minneapolis, Minnesota. His research encompasses all facets of mechanical engineering, including thermodynamics, heat transfer, fluid mechanics, and structural mechanics. Other research areas include alternative energy, traditional and nontraditional porous mediums, biomedical devices, water treatment, and other industrial applications. His teaching is focused on modeling and numerical simulation of engineering problems. He has published about 40 papers in archival journals.

Contributors

J. P. Abraham
School of Engineering
University of St. Thomas
St. Paul, Minnesota

Selin Aradag
Department of Mechanical Engineering
TOBB University of Economics and
    Technology
Ankara, Turkey

Theodore L. Bergman
Department of Mechanical Engineering
University of Kansas
Lawrence, Kansas

Amir Faghri
Department of Mechanical Engineering
University of Connecticut
Storrs, Connecticut

Mohammad Faghri
Mechanical, Industrial and Systems
    Engineering
University of Rhode Island
Kingston, Rhode Island

J. M. Gorman
Department of Mechanical Engineering
University of Minnesota
Minneapolis, Minnesota

Sadik Kakac
Department of Mechanical Engineering
TOBB University of Economics and
    Technology
Ankara, Turkey

W. J. Minkowycz
Department of Mechanical and
    Industrial Engineering
University of Illinois at Chicago
Chicago, Illinois

Ece Özkaya
Department of Mechanical Engineering
TOBB University of Economics and
    Technology
Ankara, Turkey

E. M. Sparrow
Department of Mechanical Engineering
University of Minnesota
Minneapolis, Minnesota

Bengt Sundén
Division of Heat Transfer
Department of Energy Sciences
Lund University
Lund, Sweden

Zan Wu
Division of Heat Transfer
Department of Energy Sciences
Lund University
Lund, Sweden

1 Heat Exchangers and Their Fan/Blower Partners Modeled as a Single Interacting System by Numerical Simulation

E. M. Sparrow, J. M. Gorman,
J. P. Abraham, and W. J. Minkowycz

CONTENTS

ABSTRACT: Heat exchangers usually involve two or more fluids having different temperatures. The performance of heat exchangers depends critically on the nature of the participating fluid flows. Over the years of traditional heat exchanger design, the most accounted feature of the fluid flow has been its magnitude. This is because design procedures for heat exchangers have been closely connected to fan/blower/pump curves in which the magnitude of the delivered flow is linked to the pressure rise. However, those characteristic curves do not take into account swirl, eddies, backflow, cross-sectional nonuniformities, and unusually high turbulence, almost all of which are embedded in actual fluid flows delivered to heat exchangers. The focus of this chapter is to quantitatively demonstrate the necessity of taking into account all of the characteristic features of the fluid flow that is delivered to the heat exchanger. This is accomplished by treating the fan/blower/pump and the heat exchanger as a single interactive system. In such a treatment, it is mandatory that fan rotation is fully considered to ensure that all rotation-based flow characteristics are included. The composite system, consisting of the fluid mover and the heat exchanger, is solved by numerical simulation. The fluid flows produced by this approach are a more true representation of reality.

1.1   INTRODUCTION

The goal of this chapter is to present evidence of the importance of taking into account the realistic interactions between a fluid mover and the heat exchanger, which is the recipient of the fluid flow. The underlying motivation for this work is the development of a heat transfer design methodology that is more accurate and more realistic than what is the present norm. In particular, it will be shown that the design of a heat transfer device based solely on the magnitude specification of the delivered fluid flow is insufficient because other characteristics of the flow are neglected. Depending on the nature of the fluid mover, those characteristics may include swirl, eddies, backflow, cross-sectional nonuniformities, and unusually high turbulence. The result of neglecting these factors is a design that, if implemented, may be far less efficient than what was predicted.

The foregoing discussion suggests that an improved heat-exchanger design methodology should include the realities of the delivered fluid flow. To ensure that these realities are properly characterized, the actual flow delivered by the fluid mover must be determined. In that regard, it is tempting to use the fan/blower/pump curve provided by the manufacturer of the fluid mover. The flaw in doing so is that such curves are determined after all the complexities that are normally embedded in the fan/blower output have been eliminated (i.e., the actual flow has been sanitized). Therefore, the use of fan/blower/pump curves as input to a heat transfer analysis does not necessarily lead to a good design.

In this presentation, a new methodology for heat exchanger design will be set forth followed by a succession of specific problems to illustrate its use. Ample comparisons are included to provide an indication of the accuracy improvements that can be achieved by the use of this methodology. The need to treat heat exchangers and their fluid flow providers as a single system has received very little recognition in the published literature [1–3].

1.2   FAN/BLOWER FLUID FLOW CHARACTERISTICS

The two types of fans that generally are encountered in connection with heat exchangers are axial and centrifugal. These fans have characteristics that are specific unto themselves so that it is advantageous to treat them separately. For the applications that motivated the present chapter, thermal management of electric equipment, axial fans appear to be more useful. Consequently, the focus will be directed here toward axial fans and their applications. A typical likeness of an axial fan is shown in Figure 1.1. The fan consists of a square frame or housing, blades attached to a rotating hub, and supports for the hub and its attached moving components.

The nature of the flow provided by a fan may depend on the nature of the resistance into which the discharge of the fan is blown. For illustrative purposes, the flow discharging from an axial fan into free air will be displayed by means of vector diagrams. The first pair of such diagrams is displayed on a longitudinal plane that includes the axis of the fan. In Figure 1.2a, the vectors have been normalized to have a common length. Such vectors give a picture of the directions of fluid flow. The vectors shown in Figure 1.2b have lengths that are directly proportional to the magnitude of the velocity.

The most interesting feature of Figure 1.2a is that the fan discharge is by no means confined to the downstream direction. On the contrary, there is a pronounced region of backflow extending downstream from the exit face of the fan within which a backflow prevails. This behavior can be attributed to the presence of a tornado-like rotation of the fluid that is driven by the rotating hub. The low pressure zone created by the tornadic motion sucks downstream-situated fluid forward.

Figure 1.2b shows that the highest flow rates are concentrated in the narrow vein of forward-flowing fluid that is discharged from the fan and in the back-flowing fluid that is situated just downstream of the hub of the fan. Further insights into

FIGURE 1.1   Photograph of a typical small axial fan.

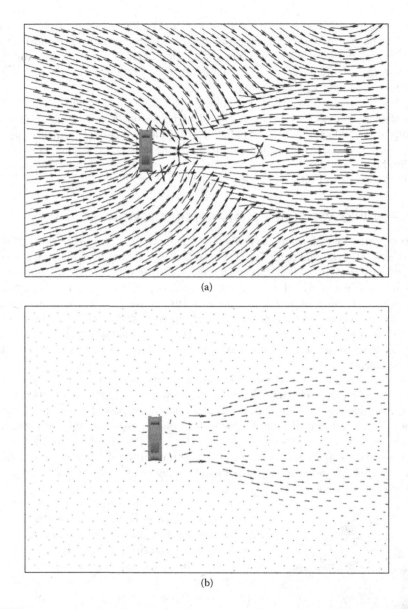

(a)

(b)

FIGURE 1.2 Vector diagrams displayed on a longitudinal plane that includes the axis of rotation: (a) normalized vectors and (b) unnormalized vectors.

the nature of the fan-produced flow are provided by vector diagrams, each of which are displayed in a selected cross section. The considered cross sections are $x/D = 0$, 1, 2, 3, 4, and 5, where x is the axial coordinate originating from the downstream face of the fan, and D is the outer diameter of the annular opening of the fan. In each cross section, both normalized and unnormalized vectors are presented side-by-side to provide unambiguous information about fluid flow direction and magnitude. The first figure in this sequence of cross sections is Figure 1.3, which corresponds to $x/D = 0$ (the fan exit cross section). The left- and right-hand parts of the figure correspond to normalized and unnormalized vectors. The normalized vectors display a general inflow of air from the surroundings toward the fan housing, whereas the flow direction within the fan proper is irregular. From the unnormalized vectors, it can be seen that the magnitude of the inflow is very small compared to the flow within the fan.

The next cross section to be visited is $x/D = 1$. A comparison of Figures 1.3 and 1.4 shows that the direction of the flow in the surroundings is hardly changed except in the near neighborhood of the fan, where a slight fluid outflow direction is seen. The magnitudes of the in-fan velocities are somewhat increased between $x/D = 0$ and 1. If focus is advanced to Figure 1.5, which corresponds to $x/D = 2$, it is seen that a coherent fluid rotation both within the fan and in its surroundings

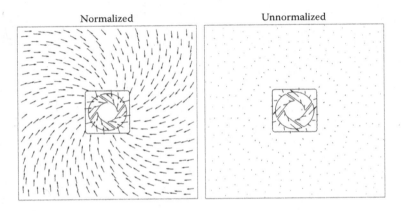

FIGURE 1.3   Flow field directions and velocity magnitudes at $x/D = 0$.

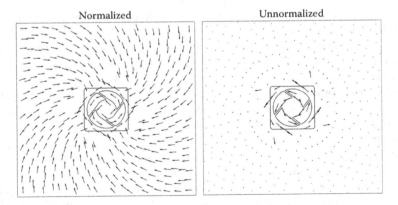

FIGURE 1.4   Flow field directions and velocity magnitudes at $x/D = 1$.

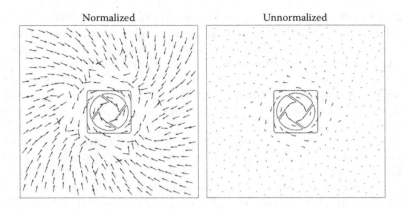

FIGURE 1.5   Flow field directions and velocity magnitudes at $x/D = 2$.

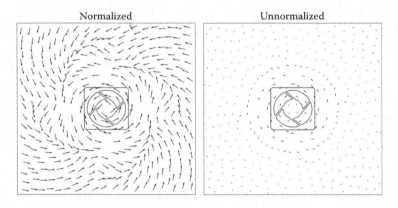

FIGURE 1.6   Flow field and velocity magnitudes at $x/D = 3$.

has now taken hold. On the other hand, the magnitude of the in-fan and near-fan fluid velocities has diminished slightly.

Figure 1.6, $x/D = 3$, shows further development and enlargement of the zone of coherent rotation of the fluid but with decreasing magnitude of the in-fan and near-fan fluid velocities.

The last two figures in this set, Figures 1.7 and 1.8, for $x/D = 4$ and 5, respectively, are virtually identical to each other. The zone of coherent rotation has stabilized, and the magnitude of the velocities appears to have reached an asymptote.

Although Figures 1.3 through 1.8 display the metamorphosis of a complex velocity field, the true complexities really are not able to be seen by means of vector velocity displays. For example, the fact that the flow is unsteady is not documented in the presented figures. Of greater significance is the heightened turbulence, which also is not seen.

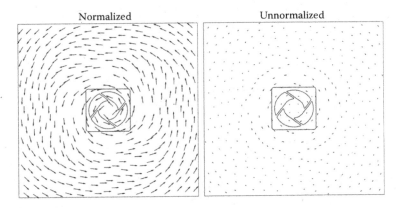

FIGURE 1.7   Flow field directions and velocity magnitudes at $x/D = 4$.

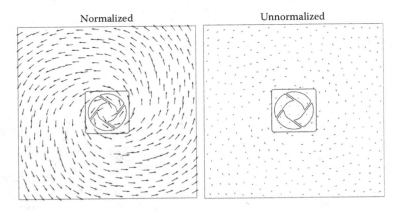

FIGURE 1.8   Flow field directions and velocity magnitudes at $x/D = 5$.

1.3   FAN/BLOWER CURVES AND APPLICATION TO DESIGN

The features that have been displayed in the preceding section, along with other complex features, have been sanitized prior to the determination of blower curves. Therefore, as noted earlier, the use of a blower curve does not represent the true nature of the flow delivered by an axial fan. Notwithstanding this, it is relevant to display how a fan/blower curve can be used as a design tool, albeit with a margin of error for the reasons just discussed.

Figure 1.9 displays fan/blower curves for two small axial fans. In general, these curves fit the general trend whereby the highest pressure rise due to the fan occurs at the lowest delivered flow rate. However, it is interesting to observe the very different shapes of the exhibited curves. Both curves display a saddle but of a very different form. The curve for Fan 2 is distinguished by a lengthy plateau that extends over half the magnitude range of the delivered flow. The sharp decrease of pressure

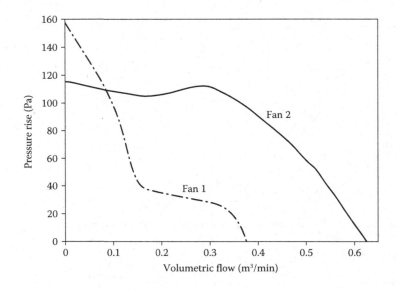

FIGURE 1.9 Operating characteristics of two axial fans (1 and 2).

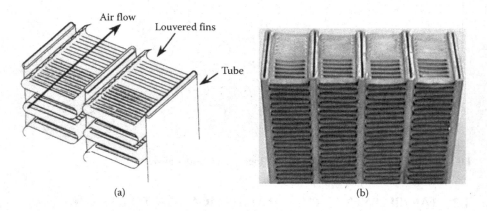

FIGURE 1.10 (a) Schematic view and (b) photograph of a louvered-fin heat exchanger.

rise occurs at large flow rates for Fan 2, whereas that sharp decrease occurs for low flow rates for Fan 1.

These fans will be used to demonstrate the relative performance of a heat exchanger equipped with flat fins and with louvered fins [4]. In general, there has been considerable interest in the use of louvered fins as witnessed by the extensive published literature. For the convenience of readers, the culled literature has been organized into two categories: numerical [5–13] and experimental [14–20]. Figure 1.10 shows both schematic and photographic views of a louvered-fin heat exchanger. The fluid flow, pressure drop, and transfer characteristics of this heat exchanger will be presented when either Fan 1 or Fan 2 is used to supply the airflow. Results for this same heat exchanger, but having flat fins, will also be presented for purposes of comparison.

There are preparatory steps that must be executed in order to enable a fan/blower curve to be utilized. Once implemented, those steps will yield the system curve. The system curve displays the pressure drop experienced by fluid passing through the system as a function of the rate of fluid flow. In order to obtain a system curve, a calculation method must be adopted. Although an estimate of the system curve can be obtained by using literature information, it is more accurate to use numerical simulations, as was done by Nguyen et al. [4]. Figure 1.11 shows the blower curves already displayed in Figure 1.9 plus the corresponding system curves.

Of particular relevance is the intersection of the system curve and the fan/blower curve. For Fan 1, the intersection in the case of the louvered fins is at a lower pressure and a lower flow rate than is the corresponding interaction with the curve for Fan 2.

The discretization of the governing differential equations and the numerical simulations was carried out using ANSYS CFX 15.0 software. Increasing the number of nodes from 14.7 to 19.6 million gave rise to variations of the heat transfer rate no greater than 0.25%. The Shear Stress Transport (SST) turbulence model was utilized for all of the investigated cases, and duplicate laminar-flow simulations were performed for the flat-fin case. Because the SST turbulence model reduces to laminar flow in situations where the flow is truly laminar, the flat-fin results obtained by use of the SST turbulence model and the laminar model were identical.

Table 1.1 displays the key outcomes of Nguyen et al. [4]. Listed there are the rates of heat transfer, the pressure drops, and the Reynolds numbers. The heat transfer rates correspond to a temperature difference of 30°C between the incoming air

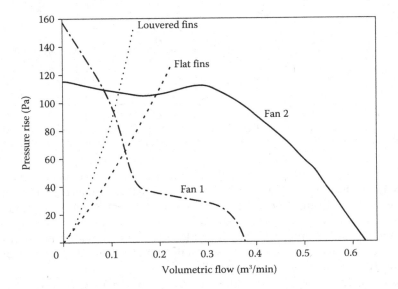

FIGURE 1.11 Intersections of the system curves for the louvered-fin heat exchanger and the flat-fin heat exchanger with the fan/blower curves for Fans 1 and 2.

TABLE 1.1

Heat Transfer and Pressure Drop Results for Louvered-Fin and Flat-Fin Heat Exchangers for Fans 1 and 2 (for a Temperature Difference of 30°C)

Variable	Fan 1		Fan 2	
	Flat	Louvered	Flat	Louvered
Heat transfer per fin (W) $\Delta T = 30°C$	0.98	0.87	1.2	0.95
Pressure drop (Pa)	65.7	91.9	108.5	111.5
Reynolds number ($\rho UD_h/\mu$)	443	483	523	805

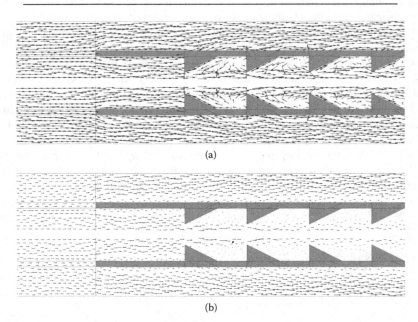

(a)

(b)

FIGURE 1.12 Flow field information for the louvered-fin situation with (a) normalized vectors showing the flow direction and (b) unnormalized vectors showing the velocity magnitudes.

and the fin base. Examination of the table reveals some unexpected outcomes. For one thing, the heat transfer rates for the so-called enhanced (louvered) surfaces are lower than those for the un-enhanced (flat) surfaces. In particular, this diminished heat transfer is more pronounced in the case of Fan 2 then for Fan 1. On the other hand, the pressure drop penalty due to the louvered fins is greatest in the case of Fan 1. The main message of Table 1.1 is that the use of enhanced surfaces does not necessarily increase the rate of heat transfer.

It is of interest to augment the information conveyed in Table 1.1 by vector diagrams that convey some insights into the nature of the flow field, and Figure 1.12 has been prepared for this purpose [4]. Figure 1.12a displays normalized vectors, whereas Figure 1.12b shows velocity magnitudes by means of unnormalized vectors. It can be seen from Figure 1.12a that recirculation zones exist in the inter-louvered spaces. Figure 1.12b shows that most of the flow bypasses those spaces. The poor performance of the louvered-fin heat exchanger may be attributed to the latter outcome.

A conclusion that follows from the results of this study is that enhancement (or disenhancement) depends as much on which fluid mover serves a heat exchanger as it does on the geometry of the exchanger in question.

The purpose of the foregoing part of the chapter is to quantify the current best practice for heat exchanger design. The insufficiency of this practice is that it omits a number of key features of the fluid flow that are delivered by the fluid mover that supplies the heat exchanger. In the remaining sections of the chapter, this omission will be remedied by modeling the actual rotation of the fluid-supply fan.

1.4 FAN WITH ROTATING BLADES: ARRAY OF FLAT-PARALLEL FINS

A favorite textbook heat transfer topic is the array of flat-parallel fins. The textbook treatment is highly introductory so that the bottom line of the treatment is a single fin. Because it is uncommon to encounter any practical heat transfer application in which there is only a single fin, the lessons learned from the typical textbook exposition are not sufficient for design. The approach to be set forth here is an in-depth treatment that would qualify as a design tool. In particular, the fan in question is treated in its full rotational activity so that all of the complexity in the delivered flow is maintained. To provide perspective on the impact of these complexities, the same problem is solved using the fan/blower curve as the description of the delivered flow.

Although it appears that the literature does not have references that relate directly to the model and the numerical solutions that are to be set forth here, it may still be useful to cite some publications that relate to the general problem of an array of parallel plain fins. There is substantial literature related to straight-parallel fins, which may be classified as: numerical [21–55], experimental [56–69], and jointly numerical and experimental [70–72].

1.4.1 PHYSICAL SITUATION

The physical situation to be dealt with is pictured schematically in Figure 1.13, which displays an array of flat-parallel fins. Air is delivered to the array by a succession of axial fans. Although fin arrays of the type exhibited in Figure 1.13 are often seen in standard heat transfer textbooks, they are shown without the fans in place.

The fin array is adiabatically shrouded from above to avoid the escape of the coolant airflow. Heat is conveyed to the fin array by means of the base surface with which the fins are integral. The fans are sized such that each fan is responsible for a particular cluster of fins. Adjacent fans do interact with each other, a situation that must be taken into account in any high-fidelity model. Such a model is displayed in Figure 1.14. The vertical faces of the outboard fins and of the upstream and downstream extensions of the solution domain were taken to be adiabatic.

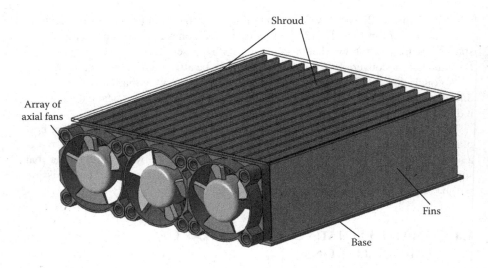

FIGURE 1.13   An array of flat, parallel fins that is shrouded above with an adiabatic surface.

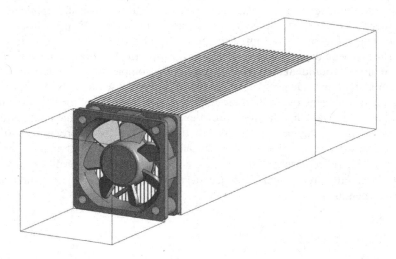

FIGURE 1.14   Diagram of the solution domain used for the numerical simulations.

For the simulation model of the problem defined in Figure 1.14, coolant air is drawn from the upstream space delineated in the figure. The coolant has a uniform incoming temperature. The fin base also has a specified uniform temperature different from that of the coolant. The rate of fluid flow provided by the fan is autoregulated by the capabilities of the fan and the resistance of the fin array. Therefore, for a given fan's revolutions per minute (RPM), there is a unique magnitude of the fan-provided fluid flow. That fluid flow magnitude gives rise to the turbulent regime. Because the magnitude of the fluid flow is not known beforehand, it cannot be specified. In this light, the boundary conditions for the velocity solution are chosen to be very weak ones so that the flow can evolve naturally.

1.4.2 Governing Equations

The conservation equations that govern the fluid flow and heat transfer include momentum, mass, and energy. In light of the fact that turbulent flow is being considered, the basic Navier–Stokes equations are replaced by the Reynolds-averaged Navier–Stokes equations, which for unsteady, constant-property, three-dimensional flow, are

$$\rho \frac{\partial u_j}{\partial t} + \rho \left(u_i \frac{\partial u_j}{\partial x_i} \right) = -\frac{\partial p}{\partial x_j} + \frac{\partial}{\partial x_i} \left((\mu + \mu_{turb}) \frac{\partial u_j}{\partial x_i} \right) \quad i = 1,2,3 \text{ and } j = 1,2,3 \quad (1.1)$$

and, for mass conservation,

$$\frac{\partial u_i}{\partial x_i} = 0 \tag{1.2}$$

The quantity μ_{turb} is called the eddy viscosity. It is, in fact, a fictitious quantity because it is not a property of the fluid. Many models have been formulated to enable the evaluation of μ_{turb}. In the present application, where attention has to be given to the swirl produced by the rotating fan blades, the sought-for turbulence model must havecredentials for being capable of dealing with the swirl. In Engdar and Klingmann's study [73], a number of turbulence models were evaluated with respect to experimental data for a situation in which swirl dominated. It was found that results predicted by the use of the SST turbulence model were in best agreement with the experimental data. Other comparisons that have supported the use of the SST model have been reported [74–76]. Those outcomes encouraged the use of the SST model for the situation of interest here. There are two turbulence parameters that are used in the SST model: the turbulence kinetic energy κ and turbulent eddy frequency ω.

The governing differential equations for these quantities are

$$\frac{\partial (\rho \kappa)}{\partial t} + \frac{\partial (\rho u_i \kappa)}{\partial x_i} = P_\kappa - \beta_1 \rho \kappa \omega + \frac{\partial}{\partial x_i} \left[\left(\mu + \frac{\mu_{turb}}{\sigma_\kappa} \right) \frac{\partial \kappa}{\partial x_i} \right] \tag{1.3}$$

$$\frac{\partial (\rho \omega)}{\partial t} + \frac{\partial (\rho u_i \omega)}{\partial x_i} = A\rho S^2 - \beta_2 e \rho \omega^2 + \frac{\partial}{\partial x_i} \left[\left(\mu + \frac{\mu_{turb}}{\sigma_\omega} \right) \frac{\partial \omega}{\partial x_i} \right]$$

$$+ 2\rho (1 - F_1) \frac{1}{\sigma_{\omega 2} \omega} \frac{\partial \kappa}{\partial x_i} \frac{\partial \omega}{\partial x_i} \tag{1.4}$$

Once κ and ω have been determined from the solution of the foregoing equations, the eddy viscosity μ_{turb} is obtained from

$$\mu_{\text{turb}} = \frac{\alpha\rho\kappa}{\max(\alpha\omega, SF_2)} \tag{1.5}$$

The First Law of Thermodynamics supplemented by the Fourier law of heat conduction is

$$\rho c_p \frac{\partial T}{\partial t} + \rho c_p \frac{\partial(u_i T)}{\partial x_i} = \frac{\partial}{\partial x_i}\left[(k + k_{\text{turb}})\frac{\partial T}{\partial x_i}\right] \tag{1.6}$$

in which the quantity k_{turb} is used to quantify the contribution of turbulence to the transfer of heat. This quantity, in common with μ_{turb}, is not a true property of the fluid. To obtain numerical values for k_{turb}, use is made of the turbulent Prandtl number Pr_{turb}.

$$Pr_{\text{turb}} = \frac{c_p\mu_{\text{turb}}}{k_{\text{turb}}} = 0.85 \tag{1.7}$$

where the numerical value 0.85 is based on comparisons of predicted heat transfer coefficients with those of experiments conducted by Churchill [77] and Kays [78].

1.4.3 NUMERICAL SIMULATION ROTATIONAL ISSUES

The rotating domain (a relatively small meshed volume surrounding the rotating components of the nondeforming fan blades and hub) uses rotating frames of reference at a specified angular velocity ω and includes both Coriolis force and centrifugal momentum terms, in addition to a rotating frame energy equation.

The additional momentum source terms are

$$S_{\text{rotation}} = -\rho\omega \times (\omega \times r) - 2\rho\omega \times u_{\theta,i} \tag{1.8}$$

where the first term is the centrifugal term, and the second term represents the Coriolis forces. In the equation, r is the position vector, and $u_{\theta,i}$ is the rotating frame velocity.

The energy equation is modified for rotation to become

$$\rho\frac{\partial I}{\partial t} + \rho\frac{\partial(u_i I)}{\partial x_i} = \frac{\partial}{\partial x_i}\left[(k + k_{\text{turb}})\frac{\partial T}{\partial x_i}\right] \tag{1.9}$$

where I is rothalpy and is defined as

$$I = h_{\text{static}} + \frac{u_i^2}{2} - \frac{(\omega R)^2}{2} \tag{1.10}$$

1.4.4 Fan/Blower-Curve-Based Analysis

In preparation for certain forthcoming comparisons, a simplified analysis based on the fan/blower curve for the fan in question will now be described. This analysis suppresses all of the complexities that accompany the flow delivered by the rotating fan to the interfin flow passages. The fan/blower curve for the fan in question (Sofasco Model d2510) is conveyed in Figure 1.15. Also appearing in that figure is the system curve corresponding to a typical flow passage in the fin array. The intersection of the fan/blower curve and the system curve provides the operating point for this model.

In this model, which avoids the realities of the rotating fan, each of the interfin flow passages behaves identically, with regard to both fluid flow and heat transfer, to any other flow passage in the fin array. Therefore, it is only necessary to solve for these processes as they occur in a single passage. The information provided by the fan/blower curve was used as input in Equations 1.1–1.7.

1.4.5 Heat Transfer Results and Discussion

The first results to be presented are those that relate to the overall rate of heat transfer for the entire fin array. Of particular relevance is the difference between the results based on the rotating fan model and those of the fan/blower-curve-based model. Another case for which results were obtained is the situation in which the fins have an infinite thermal conductivity that enables them to be isothermal. Table 1.2 conveys information about the overall heat transfer rates for the array as a whole.

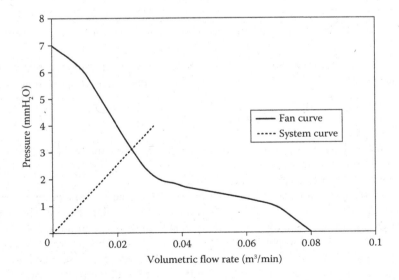

FIGURE 1.15 Fan/blower curve for Sofasco Model d2510 and the system curve corresponding to flow passing through a single passage of the fin array.

TABLE 1.2

Overall Heat Transfer Results for the Entire Fin Array Contrasting Rotating-Fan–Delivered Air with Uniform Blower-Curve–Based Delivered Airflow: $(T_{base} - T_{air,in}) = 25°C$

Heat Transfer (W)		
Rotating fan, aluminum fins	**Blower curve, aluminum fins**	**Blower curve, isothermal fins**
4.98	9.59	15.1

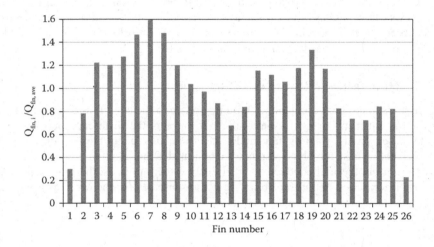

FIGURE 1.16   Ratio of the per-fin heat transfer rates to the average per-fin heat transfer rate.

The comparison between the rotating fan and fan/blower curve models reveals that the latter drastically *overestimates* the rate of heat transfer, with the outcome

$$Q_{\text{blower curve}} / Q_{\text{rotating fan}} = 1.93 \tag{1.11}$$

The size of this ratio needs explanation. It is believed that the relatively low value of Q for the rotating fan case can be attributed to the very high pressure drop and correspondingly diminished flow rate for that case.

Further study of Table 1.2 shows that the use of perfectly conducting fins does have a significant effect on the rate of heat transfer. In particular, the table shows that the isothermal fins transfer about 60% more heat than do aluminum fins.

Next, focus is shifted to per-fin heat transfer results, and this information is conveyed in Figure 1.16. This figure shows a bar graph in which each bar corresponds to a specific fin. The height of each bar displays the ratio of the heat transfer rate at the fin in question compared with the heat transfer rate averaged over all the fins. It should be noted that the outboard fins (1 and 26) are adiabatic at their outfacing surfaces, and it is this feature that is responsible for the low values of their respective heat transfer rates. These fins aside, the heat transfer ratios range from 0.7 to 1.6. The lowest among these values occurs at the fin that is situated head-on with the hub of the fan. That arrangement gives rise to blockage that, in turn, is responsible for the low heat transfer rate.

Another observation seen in the figure is a certain degree of asymmetry with respect to the central fin of the array. That asymmetry can be attributed to the direction of rotation of the fan.

1.4.6 Fluid Flow Results and Discussion

Velocity vectors will be used to illuminate patterns of fluid flow in selected channels of the fin array. The vectors lie in a plane that is parallel to the two fins that bound a channel and is situated midway between them. The first display is for the channel between Fins 6 and 7 numbered in Figure 1.16.

The vector diagram for this case is displayed in Figure 1.17a and b. Figure 1.17a shows normalized vectors for the channel in question, whereas Figure 1.17b displays unnormalized vectors. From Figure 1.17a, it is seen that there is a strong backflow that dominates the flow passage in question. Situated directly in front of the fan is a

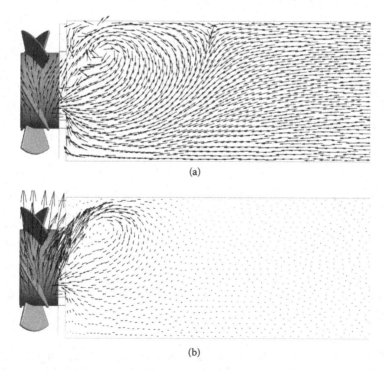

(a)

(b)

FIGURE 1.17 Vector diagrams revealing the patterns of fluid flow in selected flow passages. (a) Normalized vectors for flow passage between fins 6 and 7, (b) un-normalized vectors for flow passage between fins 6 and 7.   *(Continued)*

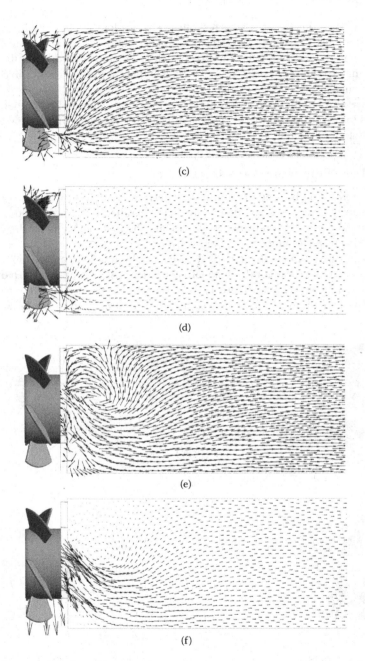

FIGURE 1.17 (Continued) Vector diagrams revealing the patterns of fluid flow in selected flow passages. (c) Normalized vectors for flow passage between fins 13 and 14, (d) un-normalized vectors for flow passage between fins 13 and 14, (e) normalized vectors for flow passage between fins 20 and 21, and (f) un-normalized vectors for flow passage between fins 20 and 21.

recirculation zone that has a backflow leg near the base of the channel and a forward flow leg near the top of the channel. A better perspective may be obtained by viewing Figure 1.17b. There, it is seen that the velocities in most of the channel are very small and that larger velocities are confined to the neighborhood near the fan. In fact, the strongest flow is concentrated in the recirculation zone.

Attention is now refocused on the flow channel between Fins 13 and 14 in Figure 1.16. This channel is identified and results are displayed in Figure 1.17c and d. Figure 1.17c shows a perfectly aligned backflow in the larger portion of the flow passage. That backflow encounters a blockage as it approaches the fan and seeks egress wherever it is able. The egress is found near the bottom of the flow passage. Figure 1.17d indicates that the backflow in the larger portion of the channel is of moderate velocity; only in the near neighborhood of where the fluid leaves the passage are larger velocities encountered.

The third flow passage to be explored is that between Fins 20 and 21 in Figure 1.17e and f, respectively. Inspection of Figure 1.17e discloses a strong forward flow in the passage proper. However, there is a more complicated zone of flow in the near neighborhood of the fan. A recirculation zone is seen to exist near the upper part of the fan. This recirculation zone forces the flow leaving the fan to follow a downward trajectory. Figure 1.17f provides further information. It shows that the highest velocities are in the region of the downward trajectory. In the larger expanse of the flow passage, the forward flow is of a moderate velocity.

1.4.7 Retrospective View of the Flat-Fin Array Results

It has been demonstrated in the foregoing that the performance of a heat exchanger is strongly impacted by the nature of the flow delivered to it by a specific fan/blower. It is clear that the performance of a heat exchanger is actually the result of the performance of a *system* that consists of the exchanger and the fan/blower. This outcome has important practical ramifications.

A second outcome of the foregoing presentation is that predictions based on the use of a fan/blower curve model as a simplification of the actual rotating fan model is not likely to give accurate results. This is because the fan/blower curve is confined to conveying the magnitude of fluid flow without actually providing information about the true nature of the flow. The nature of the flow is characterized by swirl, eddies, backflow, cross-sectional nonuniformities, and unusually high turbulence in addition to the magnitude of the flow. The sanitization of the delivered flow by the fan/blower curve is the factor that gives rise to errors that are encountered when the fan/blower curve is used as the basis of design.

1.5 FAN WITH ROTATING BLADES: APPLICATION TO A DNA THERMOCYCLING DEVICE

Thermal processes play a major role both in biomedical therapeutic devices and in biomedical diagnostic instrumentation. Here, attention is focused on a heat-transfer-based device that may be regarded as biomedical diagnostic instrumentation. The device in question, commonly designated as a *thermocycler* or a *PCR machine*, processes DNA samples according to a protocol that subjects samples to cyclic temperature variations between prescribed upper and lower bounds. The cycling is accomplished by the successive use of heating and cooling means. Heating is accomplished by thermoelectric chips, electrical resistive heating, convection, and induction. On the other hand, cooling may be achieved by thermoelectric means commonly supported by a heat exchanger referred to as a heat sink in the PCR literature. The heat sink in question corresponds to a real-world product so that it is geometrically more complex than heat exchangers encountered in academic situations.

There is a rich literature dealing with PCR machines [79–83]. For the most part, that literature is less focused on heat transfer issues than on other aspects of the device. In particular, in none of the literature surveyed was there any significant investigation of the heat sink characteristics. Here, the heat sink will be investigated in detail by numerical simulation.

It appears that the heat sink dealt with here differs from what has been considered elsewhere.

1.5.1 PHYSICAL SITUATION

A version of a DNA thermocycler (PCR device) is displayed in Figure 1.18. It can be seen from the figure that the coolant air is drawn into the machine through a narrow gap between its bottom and the table on which the device is situated. Although the innards of the machine are not visible in this diagram, it can be envisioned that the internal air flow path is quite complex because the air exits in a direction that is very different from its inflow direction. A thick, temperature-controlled aluminum block sits atop the main casing of the machine. The upper surface of the block is perforated with an array of regularly shaped holes to accommodate a collection of DNA-filled vials.

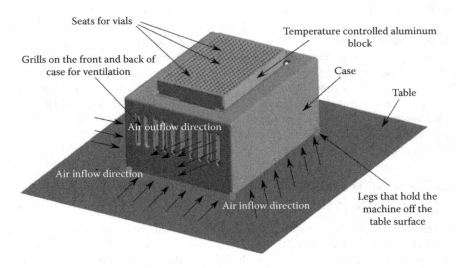

FIGURE 1.18 Pictorial view of a thermocycler in a fully assembled condition.

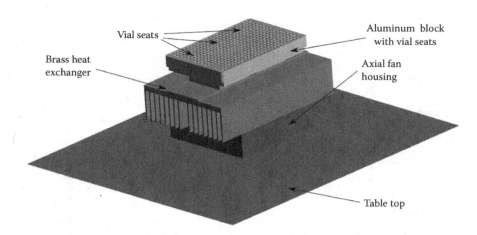

FIGURE 1.19 Pictorial view of a thermocycler with its main casing removed.

The removal of the main casing of the machine is depicted in Figure 1.19. That view shows the housing of an axial fan above which a fin array is in place. The fan is the means by which air is drawn into the fin array from the surroundings. It is noteworthy, however, that the air experiences a right-angle turn from its original inflow direction as it negotiates the fan inlet cross section. Because the fan blows upward, the air enters the fin array through a plane that encompasses the fin tips. This location of air ingress differs significantly from that which was considered in Section 1.4 and displayed in Figure 1.14. The major difference in the mode of ingress relevant to the present problem and that of the preceding one suggests that the fluid flow for the two situations will vary greatly.

Another geometrical feature evident in Figure 1.19 but that was hidden in the preceding figure is a slab-like form situated between the fin base and the aluminum block. That slab represents a thermoelectric chip that functions to cool or heat the block. The thermoelectric chip has two principle faces—one of which is maintained at a below-ambient temperature while the other has a temperature well above ambient. The cool, fan-driven air impinges on the hot face of the chip and cools it. Figure 1.20 reinforces the view that is seen in Figure 1.19 and shows more clearly the thermoelectric chip, an intervening aluminum plate, and the direction of the air flow as it enters the interfin spaces.

1.5.2 Governing Equations

The focus of the work is the solution of a convective heat transfer problem. However, a necessary requisite is the solution for the velocity field. The velocity field for the present situation is different and more complicated than that for the problem dealt with in Section 1.4. The major difference in the two cases is the manner in which the airflow enters the fin array. The direction of air ingress outlined in Section 1.4 is from one end, as can be seen in Figure 1.14; once air is within the fin array, the pattern of air flow is more or less longitudinal. In contrast, Figure 1.20 shows air

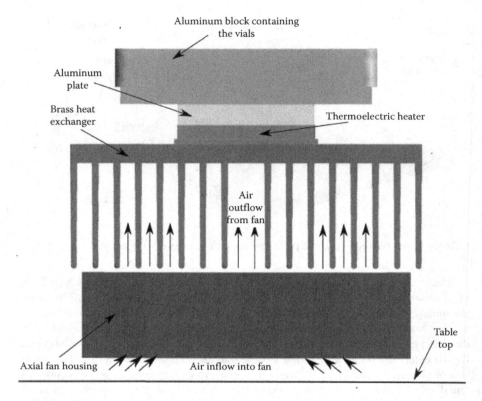

FIGURE 1.20 Side view of the internal components of the thermocycler.

entering the interfin spaces through a plane that encompasses the fin tips. This air penetrates to the fin base where it splatters and becomes a longitudinal flow, but it also attempts to exit through the part of the fin-tip plane that is not blocked by the fan. Figure 1.21 shows the open space in the fin-tip plane that is not blocked by the fan. In light of the foregoing discussion, the solution of the DNA thermocycler is expected to be much different from that obtained in Section 1.4.

The governing equations for the problem now being considered are the same as those for the preceding problem. The primary difference in the corresponding numerical simulations is in where the boundary conditions are applied. In particular, the inlet and exit fluid flow boundary conditions have to take into account the specific locations of the inlets and the exits. The SST turbulence model is used once again.

1.5.3 Heat Transfer Results and Discussion

The overall area-averaged heat transfer coefficient will be the first outcome to be discussed. That quantity was determined from the overall rate of heat transfer divided by the total surface area of the fins and by the temperature difference between the base surface and the ambient air temperature. The final result is that h_{ave} = 4.07 W/m^2 °C.

Axial fan housing   Axial fan blade

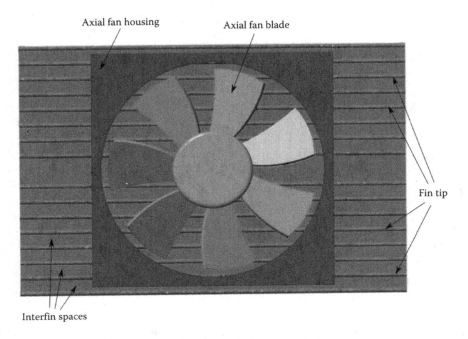

Fin tip

Interfin spaces

FIGURE 1.21   Juxtaposition of the fan and the plane of the fin array.

This value is surprisingly low and requires some discussion. Careful examination of the pattern of fluid flow by means of vector diagrams, to be displayed shortly, revealed that some of the heated air exiting from the fin array was sucked back into the fan inlet and was thereby returned to the fin array. This means, in reality, that the fins were washed by air of a higher temperature then that of the ambient. As a consequence, the actual temperature difference driving the heat transfer is smaller than the difference between the base and the ambient temperatures. Now, with the reason for the poor heat transfer performance understood by means of the velocity solution, the information needed to improve the performance is at hand. Clearly, the geometry of the airflow path has to be altered so that the heated air exiting the fin spaces is not able to return to the fan inlet.

Attention will now be turned to the per-fin heat transfer results. The rate of heat transfer at each face of a given fin was determined and the results plotted in Figure 1.22. The vertical axis of the figure is the ratio of the rate of heat transfer at each face of each fin divided by the array-averaged value for each face of each fin. The fin numbers are distributed along the horizontal axis. The left-hand face of each fin is denoted as the A-face, and the right-hand face of each fin is the B-face. An overall inspection of the figure indicates that, aside from the neighborhood of the outboard fins, there is not a consistent pattern of difference between the A and B faces. With regard to the outboard fins, the outfacing surface of each such fin experiences higher rates of heat transfer than does the in-facing surface of that fin.

Figure 1.23 provides a different viewpoint of the results of Figure 1.22. In the new figure, the heat transfer rates for the A and B sides of each fin are added and

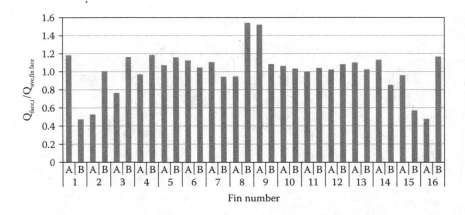

FIGURE 1.22 Ratio of the heat transfer rate per fin face to the average fin rate of heat transfer.

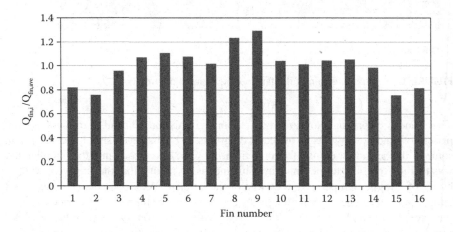

FIGURE 1.23 Ratio of heat transfer rate per fin to the average per-fin heat transfer rate.

then plotted as a single per-fin value. It is seen from Figure 1.23 that 10 of the 16 fins are providing a heat transfer rate that is approximately equal to the array-averaged rate. The deviations from this pattern occur near the outboard ends of the fin array and also at the center of the array where the interfin spacing is larger than that for the other fins. This nonuniformity in the spacing can readily be seen in Figure 1.21.

Further depth of exposition will reveal other insights into the nature of the heat transfer. To achieve the desired further depth, a color (gray tone) contour diagram of the distribution of the heat flux on each of the two faces of each fin was extracted from the numerical solutions. Sample results from among the totality of the available color contour diagrams will now be presented in Figures 1.24, 1.25, and 1.26. Figure 1.24 shows a pair of color contour diagrams (in gray tones) for fin surfaces 3B and 4A. These surfaces bound the third flow passage as viewed from the left edge of

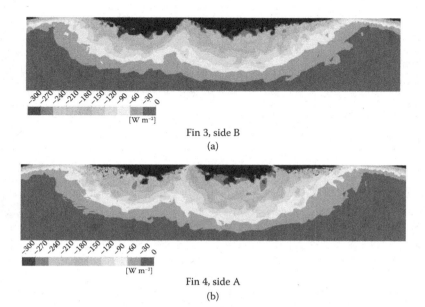

Fin 3, side B

(a)

Fin 4, side A

(b)

FIGURE 1.24　Color (gray tone) contour diagrams showing the spatial variation of the heat flux on the surfaces of Fins 3B and 4A. These fins bound the third flow passage as viewed from the left edge of the fin array.

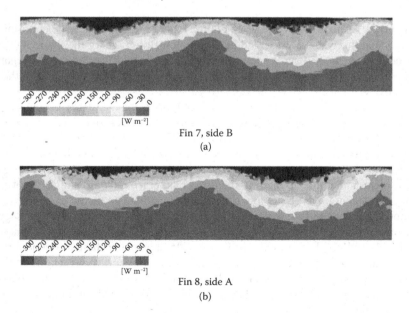

Fin 7, side B

(a)

Fin 8, side A

(b)

FIGURE 1.25　Color (gray tone) contour diagrams showing the spatial variation of the heat flux on the surfaces of Fins 7B and 8A. These fins bound the seventh flow passage as viewed from the left edge of the fin array.

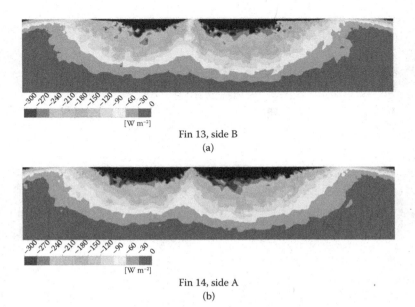

[W m⁻²]

Fin 13, side B

(a)

[W m⁻²]

Fin 14, side A

(b)

FIGURE 1.26   Color (gray tone) contour diagrams showing the spatial variation of the heat flux on the surfaces of Fins 13B and 14A. These fins bound the thirteenth flow passage as viewed from the left edge of the fin array.

the fin array. Each of the two parts of Figure 1.24 has a gray tone scale coordinated with the heat flux value represented by each color. Both parts of the figure have the same color scale. Note that the color scale expresses the heat flux values as a negative number that, in the convention used here, indicates that the heat flow is from the fin to the airflow.

The upper edge of each fin face displayed in Figures 1.24–1.26 corresponds to the location of the tips, whereas the lower edge of the face corresponds to the junction of the fin and the base. Inspection of the two parts of Figure 1.24 reveals a color pattern that is virtually the same. The highest heat fluxes occur in the neighborhood of the fin tips, and the lowest fluxes occur in the neighborhood of the fin base. This observation suggests that the air entering the flow channel from the tip end does not have sufficient momentum to propel it to a strong impingement on the base.

Information of the type conveyed by Figure 1.24 for the third flow passage is presented in Figure 1.25 for the seventh flow passage. This flow passage is bounded by Fins 7B and 8A. Examination of Figure 1.25 shows that the two bounding fins exhibit heat flux distributions that are only slightly different from each other. Once again, in concert with Figure 1.24, the highest heat flux occurs in a narrow band in the neighborhood of the fin tips. There is a wider band of low heat flux in the neighborhood of the fin base.

The final figure in this sequence, Figure 1.26, exhibits information on the surfaces of Fins 13B and 14A. These surfaces bound the thirteenth flow passage. It is seen from this figure that, once again, the two fins that bound a flow passage experience very similar heat flux patterns. In fact, an overview of Figures 1.24–1.26 indicates that there is a consistent heat flux pattern for all of these situations.

1.5.4   Fluid Flow Results and Discussion

Quantitative information about the pattern of fluid flow will be conveyed by two approaches. The first approach will exhibit the rates and locations of fluid outflow from the interfin channels, whereas the second will make use of vector diagrams. Figure 1.27 displays the rates of fluid flow leaving the interfin channels via four different outflow directions. Two of these directions are the respective outboard ends of each passage. These ends are reached by airflow that adopts a longitudinal flow direction after being incident on the fin base. The other two outflow directions are through the parts of the plane that contain the fin tips that are not blocked by the fan. To clearly identify these directions, a code based on gray tones is situated at the lower edge of the figure.

Inspection of Figure 1.27 reveals that, for the most part, the majority of the outflow occurs at the outboard ends of the respective channels. The only exceptions to this observation are the outboard edge Channels 1 and 15, where chaos generally prevails, and the center channel (Channel 8).

Vector diagrams are now presented to provide further insights into the patterns of fluid flow. The first pair of diagrams is exhibited in Figure 1.28a and b. Figure 1.28a shows normalized vectors and thereby displays flow directions. Magnitudes of the fluid flow are seen in Figure 1.28b. The view conveyed in Figure 1.28 is an end view, which is similar to that already shown in Figure 1.20.

Observation of Figure 1.28a shows that the flow entering the interfin spaces is by no means unidirectional. Of particular interest are the partial blockages adjacent to the two outboard flow passages. The fluid that seeps through these blockages is

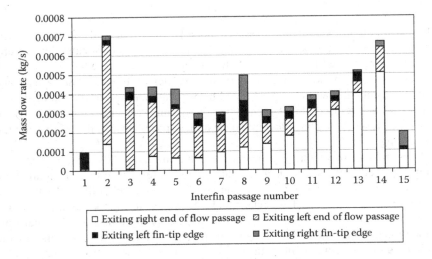

FIGURE 1.27   Fluid outflow directions and magnitudes for the respective flow channels.

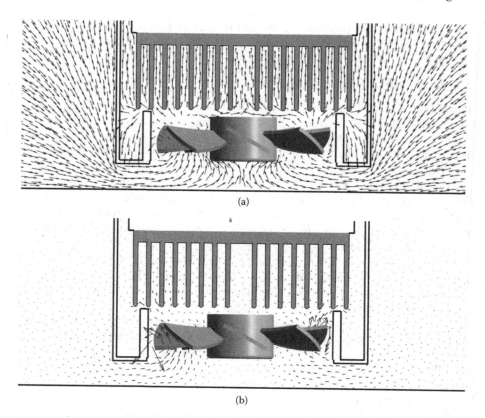

FIGURE 1.28 Vector diagrams in a plane cut through the axis of the fan and perpendicular to the longitudinal direction of flow through the fin passages: (a) normalized vectors and (b) unnormalized vectors.

subdivided among the outboard passages and a corridor that leads out of the array. It would appear that the outflow is ultimately sucked back into the fan inlet, thereby providing a preheated flow to the interfin passages. This is a negative factor with respect to the magnitude of the heat transfer coefficient. Figure 1.28b reveals that flow of substantial magnitude is confined to the neighborhood of the fan inlet. The aforementioned re-used flow is seen to be of very small velocity. Figure 1.28a and b shows the presence of a backflow in the clearance space between the fin tips and the fan housing.

A second pair of vector diagrams is presented in Figure 1.29a and b. The viewpoint displayed in this figure is at a right angle with respect to that of Figure 1.28. This viewpoint shows the flow passing into the center channel of the array, with one of the bounding fin surfaces as background. The flow entering the passage is shown in Figure 1.29a, impacting on the fin base and morphing into a longitudinal flow in the direction of the outboard ends of the passage. Here again, it appears that the outflowing fluid is recycled and returns to the fan inlet. Some vestige of that recycled flow can be seen in Figure 1.29b. This observation suggests that an increase in the

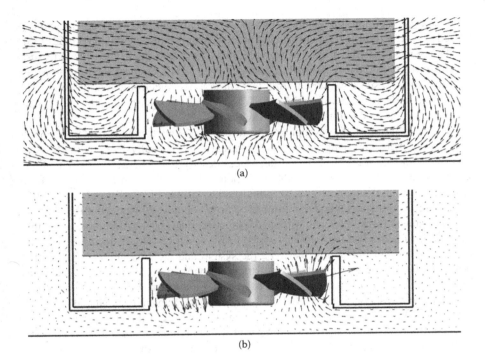

(a)

(b)

FIGURE 1.29 Vector diagrams in a plane cut through the axis of the fan and parallel to the longitudinal direction of flow through the fin passages: (a) normalized vectors and (b) unnormalized vectors.

heat transfer performance of the system could be achieved by rerouting the direction of the flow exiting from the fin array.

1.5.5 RETROSPECTIVE VIEW OF THE DNA SAMPLING DEVICE SIMULATIONS

The results of the simulations performed for the DNA sampling device illuminate the value of using the numerical simulation approach as a design tool. These results have pinpointed the reason for the unexpectedly low value of the heat transfer coefficient. The low value is the result of the recycling of fluid that is heated by its passage through the fin array. With this attribution, remedial action can be taken.

1.6 FAN WITH ROTATING BLADES: PIN-FIN HEAT SINK

1.6.1 PHYSICAL SITUATION

Still another example of the use of numerical simulation methodology as a high-fidelity design tool is the analysis of a pin-fin heat sink. The single-system approach is the novel feature of the present work. The importance of pin fins as a heat exchanger methodology is documented by a very extensive published literature. This literature may be organized into three categories: numerically based [84–113], experimentally based [114–151], and encompassing both numerical and experimental parts [152–155].

The present model is shown in Figure 1.30 as a single system. The special feature of this design problem is the full accounting of the interaction of the fan–fin array system with the surrounding environment. In practice, electronic equipment would be affixed to the rear face of the fin base. The base serves to distribute the heat generated by the equipment to the pin-fin array.

The accounting of the complex flow drawn into and through the fan, through the pin-fin array, emptying into the surroundings, and partially recirculated back to the inlet requires a highly detailed numerical description. The governing equations are the same as those already stated in Equations 1.1–1.10. The transformation of these partial differential equations was accomplished by means of ANSYS CFX 14.0 software. For accurate solutions, the number of nodes deployed throughout the solution domain ranged from 11.5 to 14 million. Mesh independence studies demonstrated that this range of nodes yielded solutions whose heat transfer results are accurate to 0.2%.

The significant interactions between the system and the environment require that special attention be given to the size of the solution domain. Figure 1.31a and b has been prepared to display the outlines of the solution domain. In Figure 1.31a, a one-quarter, top-down plan view of the system and its interacting environment, all of which are included in the solution domain, is shown. The choice of the one-quarter portion reflects the intrinsic geometric symmetry of the situation. Figure 1.31b, in contrast, is an elevation view. With regard to Figure 1.31a, it is clear that the solution domain

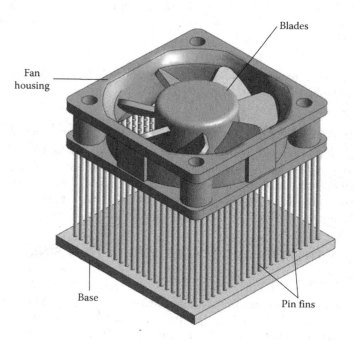

FIGURE 1.30 Pictorial view of a pin-fin array with an air-delivery fan in place.

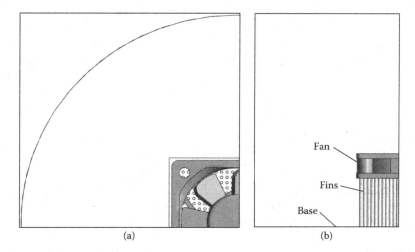

FIGURE 1.31 Diagrams showing the extent of the solution domain in both (a) the plan view and (b) the elevation view.

extends well beyond the lateral dimensions of the array, thereby enabling the flow exiting the array to circulate freely. Figure 1.31b shows that there is a large space above the fan inlet plane within which natural fluid motions can occur without restrictions.

Figure 1.31 enables a convenient description of the boundary conditions. In Figure 1.31a, the horizontal and vertical straight lines are periodic boundaries, reflecting the rotational motion of the fan blades. The curved boundary is an *opening* through which fluid can either pass inward or outward. In the elevation view (Figure 131.b), the left-hand vertical line represents an opening, as does the topmost horizontal line. The right-hand vertical line is a periodic boundary, while the lowermost horizontal line demarks the surface of the fin base. The pressure at all openings is the ambient value, and at all solid surfaces, both the perpendicular and tangential velocity components are zero.

The solution agenda for the aforementioned physical situation consists of three parts. The first part is a fully encompassing numerical modeling of the actual physical situation, including fan rotation, and the subsequent computer implementation of that model. The second part makes use of the fan/blower curve for the fan in question and interprets the information from that curve as a steady, uniform flow entering the plane of the fin tips. A third variant uses the mass flow rate actually delivered to the fin-tip plane by the rotating fan and interprets that flow as steady and uniform. For all three of these models, numerical solutions were based on Equations 1.1–1.10.

1.6.2 FAN/BLOWER-CURVE-BASED ANALYSIS

Figure 1.32 is a display of the fan/blower curve for the fan in question (Sofasco fan Model d5015, 5600 RPM). In addition to the fan curve, there are two system curves. One of these curves corresponds to an array of fins 15 mm in height, and the other is for an array of fins 25 mm in height. The fins are 0.5 mm in diameter and are made

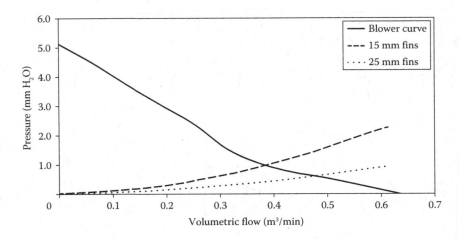

FIGURE 1.32 Fan/blower curve for a Sofasco fan, Model d5015 (5600 RPM), and the system curves for pin-fin arrays populated by fins 15 and 25 mm in height, respectively, and 0.5 mm in diameter.

of aluminum. The determination of each system curve involved numerical simulations based on selecting a number of flow rates and interpreting each flow rate as corresponding to a uniform velocity distribution crossing a plane that encompasses the fin tips. For each such selected flow rate, the pressure drop through the system was obtained by numerical simulation, and the knowledge of both flow rate and pressure drop enables a point on the system curve to be obtained. The operating point for each of the two fin arrays corresponds to the intersection of the fan/blower curve and the appropriate system curve.

It may appear from an inspection of Figure 1.32 that the intersection points are not consistent with physical intuition; in particular, it might have been expected that the taller fin array might display a larger pressure drop than that for the shorter fins. However, this seeming inconsistency can be resolved by noting that the taller fins allow air to leak out laterally in preference for full penetration to the fin base.

1.6.3 HEAT TRANSFER RESULTS AND DISCUSSION

The first result to be presented is the overall rate of heat transferred from the entire fin array to the coolant air. This information is conveyed in Table 1.3 for the case of a 25°C temperature difference between the fin base and the coolant air inlet temperature. The table displays results for the three investigated models: the fully complete rotating fan model, the blower curve model, and the matched flow-rate model whereby the flow rate obtained from the rotating fan solution is morphed into a uniform steady flow. Inspection of the table reveals that the blower curve model greatly overestimates the overall heat transfer rate by a factor on the order of 40% or more. This outcome can be attributed to the high fluid resistance that works to diminish the rate of coolant flow that the rotating fan delivers to the fin array.

TABLE 1.3
Overall Heat Transfer Rates in Watts for the
Fin Array for $\Delta T = 25°C$

Model	15 mm Fin Height	25 mm Fin Height
Blower curve	40.6	53.9
Rotating fan	26.8	33.8
Matched flows	22.0	21.3

The fan/blower curve model is not impacted by the flow-diminishing effect of this resistance. Clearly, a design based on a blower curve model would not be not be suitable for a practical application.

Also of interest in the table is the performance of the matched flow-rate model. That model is based on taking the flow rate delivered to the array by the rotating fan and spreading that flow rate into a uniform steady velocity. The heat transfer performance of that model is strongly diminished by the low value of the flow rate but is not assisted by the mixing motions created by the fan rotation.

Attention is next focused on the per-fin heat transfer rates. All told, the full array consisted of 676 individual pin fins. If geometric quarter symmetry is invoked, there remain 169 pin fins each having a unique thermal performance. The per-fin thermal performance results will be conveyed by the ratio $Q_{fin,i}/Q_{fin,ave}$, in which $Q_{fin,i}$ is the per-fin rate of heat transfer for fin i, and $Q_{fin,ave}$ is the rate of heat transfer averaged over all of the fins. These ratios are conveyed in the array of circles that constitute Figure 1.33, where each circle corresponds to one of the fins in the array. Because the array is two-dimensional, numbers that serve as the coordinates are deployed along the bottom and the right side. The exact center of the array is at the center of the fin (1, 1). The uppermost boundary of the array is open to the external environment as is the leftmost boundary. The figure corresponds to the model based on the blower curve and is for the fin height of 15 mm.

An overall examination of the figure shows per-fin heat transfer performance that is relatively uniform over the entire span of the array, with the per-fin heat transfer ratios ranging from 0.49 to 1.22. The lowest values occur along the vertical line adjacent to the right-hand edge and along the horizontal line adjacent to the lower edge of the figure. In addition, there is a diagonal extending from the lower right to the upper left where a series of somewhat lower ratios occur. Aside from these locations, the heat transfer ratios are generally greater than one.

Consideration will next be extended to the results that correspond to the rotating fan model, and this information is presented in Figure 1.34 in a format that is the same as that used for Figure 1.33. Format aside, the two figures display substantially different outcomes. The main difference between the results is the very large variation in the heat transfer performance for the array that is serviced by the rotating fan in contrast to the array that receives fan/blower curve air. For the former case, the ratio $Q_{fin,i}/Q_{fin,ave}$ ranges from 0.02 to 2.49. The zone adjacent to the center of the array is generally a region of low values of the ratio.

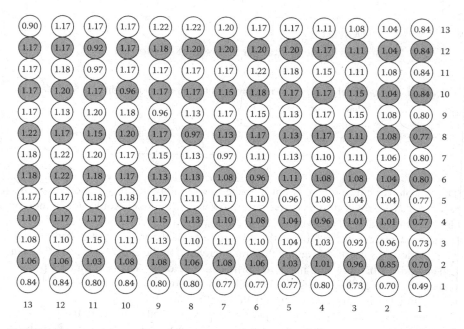

FIGURE 1.33 Per-fin heat transfer performance for the array of 15-mm-high fins serviced by the fan/blower–curve model. The displayed values correspond to the ratio $Q_{fin,i}/Q_{fin,ave}$.

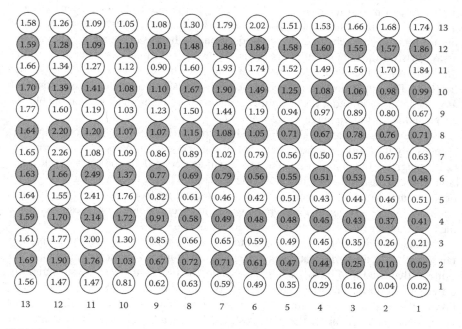

FIGURE 1.34 Per-fin heat transfer performance for the array of 15-mm-high fins serviced by the rotating fan model. The displayed values correspond to the ratio $Q_{fin,i}/Q_{fin,ave}$.

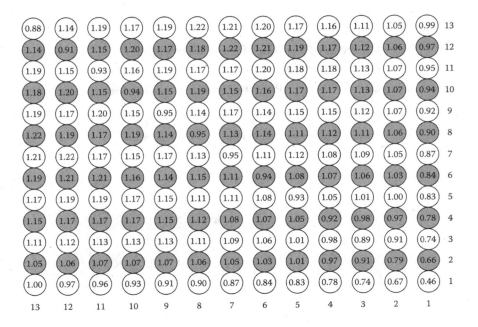

FIGURE 1.35 Per-fin heat transfer performance for the array of 15-mm-high fins serviced by the matched mass flow model. The displayed values correspond to the ratio $Q_{fin,i}/Q_{fin,ave}$.

The final figure in this set is Figure 1.35. That figure corresponds to a uniform velocity inflow based on the mass flow rate delivered to the array by the rotating fan. An overall observation of Figure 1.35 reveals that the displayed values of $Q_{fin,i}/Q_{fin,ave}$ are very similar to those seen in Figure 1.33. This degree of similarity can be attributed to the fact that both the blower curve case (Figure 1.33) and the matched mass flow rate case (Figure 1.35) are based on a uniform inlet velocity. The respective velocity magnitudes for the two cases are quite different, so that the respective values of $Q_{fin,ave}$ are also different. However, because of the linearity of the heat transfer problem, the ratios are independent of the magnitude of the inlet velocity.

1.6.4 FLUID FLOW RESULTS AND DISCUSSION

The heat transfer results will now be augmented by fluid flow information. The first pieces of information, conveyed in Table 1.4, are the volumetric flow rates associated with the different models. Clearly, the flow provided by the fan/blower curve model is far greater than that for the rotating fan model. This outcome can be attributed to the greater flow resistance that is created by the interaction of the rotating fan with the blockage caused by the pin-fin array.

Vector diagrams, presented in Figures 1.36–1.40, provide detailed information on the patterns of fluid flow. Each figure consists of two parts, (a) and (b), conveying normalized and unnormalized vectors, respectively. Figures 1.36 and 1.37

TABLE 1.4

Volumetric Flow Rates for the Investigated Physical Situations (m³/min)

Case	15 mm Fin Height	25 mm Fin Height
Rotating fan	0.108	0.0993
Blower curve	0.390	0.480
Matched flow rates	0.108	0.0993

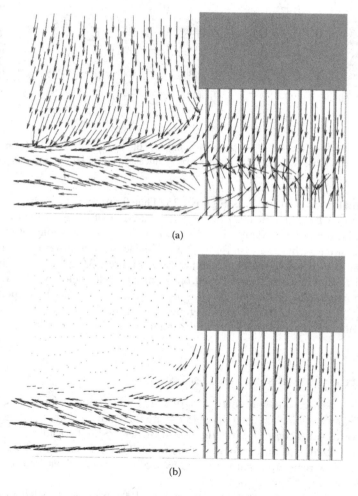

(a)

(b)

FIGURE 1.36 Vectors for the fan/blower curve model for 15-mm-high fins on a vertical plane located at the center of the fin array, (a) normalized vectors and (b) unnormalized vectors.

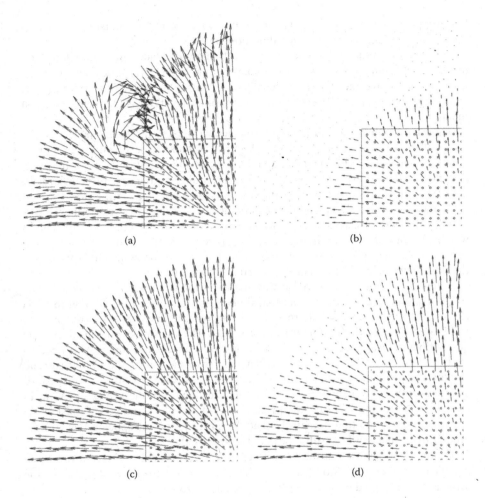

FIGURE 1.37 Vectors for the fan/blower curve model for 15-mm-high fins on horizontal planes located 5 and 10 mm above the fin base: (a) normalized vectors and (b) unnormalized vectors at the plane 5 mm above the base; (c) normalized vectors and (d) unnormalized vectors at the plane 10 mm above the base.

correspond to the fan/blower curve model. They are presented before those of the rotating fan model because they are simpler and easier to comprehend.

Figure 1.36 displays vectors in a vertical plane that passes through the center of the fin array. Figure 1.36a shows flow directions. The flow enters the array at the fin tips with a velocity that is uniform and vertical. As the flow proceeds vertically downward, it is vectored off to the left by the recognition that the exit is in that direction. The exiting flow persists about two-thirds of the way to the fin base but appears to have insufficient momentum to persist to the surface of the base. Instead, an inflow is seen to occur—in all likelihood because the pressure in the interior of the array has dropped below the ambient. The down flow to the left of and outside the array is due to entrainment activated by the fluid flowing out of the array.

Velocity magnitudes that correspond to the flow displaced in Figure 1.36a are displayed in Figure 1.36b. There, it is seen that the highest velocities belong to the flow that has exited the fin array. The velocities within the array proper are moderately large, but only small vectors are seen at other locations aside from those just mentioned.

Flow field information in horizontal planes parallel to the base surface for the fan/blower curve model is presented in Figure 1.37a and b for a plane that is 5 mm above the base and in Figure 1.37c and d for a plane that is 10 mm above the base. The flow patterns revealed in these figures are quite regular. There is an orderly flow from the interior of the array into the environment. The exiting flow has two paths into the environment, and the vectors internal to the array arrange themselves to enjoy the path of least resistance to the respective exits. The only unique feature worthy of note can be seen in Figure 1.37a. There, the outflows through the respective exits interfere with each other and create a recirculation zone. Other than that, the flow patterns in the two planes are virtually the same.

Attention is now turned to the flow patterns for the rotating fan model. The starting point is the diagrams in a vertical plane as seen in Figure 1.38. Figure 1.38a clearly displays an unruly pattern that greatly contrasts with the pattern displayed in Figure 1.36a for the fan/blower curve model. The fan draws air into it, but the air motions that are seen in the interior of the array are not logically connected to the inflow direction. A large eddy stands just to the left of the inflow current, and there appears to be incompatibility between the directions of these two streams. This incompatibility is due to the insufficient resolution of the vector diagram. In reality, there are small recirculation zones between the two streams. For most of the boundary of the solution domain beyond the array proper, it appears that inflows predominate. The outflow path is seen at the lower left.

Figure 1.38b shows that the largest velocities are existent in the inflow current and the outflow current. Moderate velocities are present in the internal portion of the array and in a limited part of the large recirculation zone.

Vectors in planes parallel to the base are displayed in Figures 1.39 and 1.40 for plane locations that are, respectively, 5 and 10 mm above the base. Figure 1.39a and b shows, respectively, the directions and magnitudes of the flow. The flow directions shown in Figure 1.39a are quite regular in far field away from the array. However, within the array and its immediate surroundings, it appears that there is a pattern, suggesting fluid rotation. The velocity magnitudes seen in Figure 1.39b are strongest along an isolated outflow path. Moderate velocities are found in more dispersed outflow directions.

The last figure in this sequence, Figure 1.40, displays vectors in a plane that is 10 mm above the base. In Figure 1.40a, the influence of the fan rotation is clearly in evidence in the environment that is closest to the fin array. In the far field, the vectors are deployed in a relatively regular manner. With regard to velocity magnitude, the highest velocities seen in Figure 1.40b are in the rotational flow in the near environment of the array.

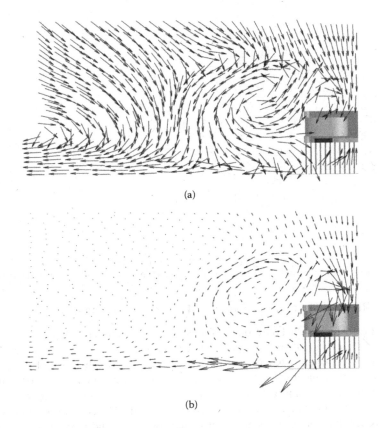

(a)

(b)

FIGURE 1.38 Vectors for rotating-fan-driven flow for 15-mm-high fins on a vertical plane located at the center of the fin array: (a) normalized vectors and (b) unnormalized vectors.

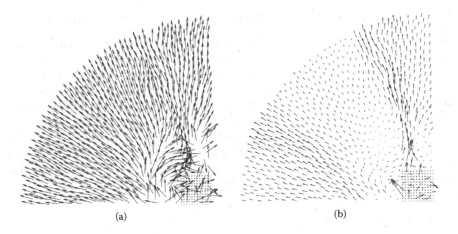

(a)   (b)

FIGURE 1.39 Vectors for fan-driven flow for 15-mm-high fins on a horizontal plane located 5 mm from the fin base, (a) normalized vectors and (b) unnormalized vectors.

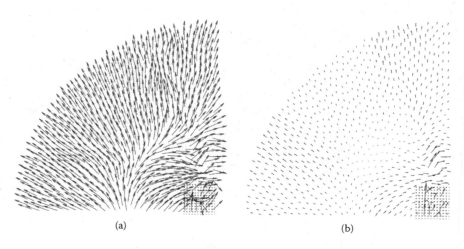

(a)   (b)

FIGURE 1.40 Vectors for rotating-fan-driven flow for 15-mm-high fins on a horizontal plane located 10 mm above the fin base, (a) normalized vectors and (b) unnormalized vectors.

1.6.5   Retrospective View of the Pin-Fin Heat Sink Investigation

The heat transfer results based on the fan/blower curve model are seen to have overestimated the more realistic results from the rotating fan model. This pattern of overestimation corroborates what was found in Section 1.4 for the case of flat-parallel fins. It may be conjectured that this pattern may be universal. This conjecture is based on the fact that fan/blower curve information is more sanitized compared with the nature of the flow that is actually delivered by a fan. The actual flow may include swirl, eddies, backflow, cross-sectional nonuniformities, and unusually high turbulence. None of these characteristics are captured by the fan/blower curve.

1.7   FAN WITH ROTATING BLADES: AIR BLOWN INTO PIPE INLET

The final application to be addressed in this chapter is geometrically less sophisticated than those dealt with earlier. It is the case of a straight, round pipe with an axial fan blowing air directly into the inlet cross section of the pipe. The swirl imparted to the flow by the rotation of the fan blades is carried downstream by the flow's axial component. The literature contains many references to the use of swirl-producing inserts in pipe flows. An often-used swirl-producing device is a twisted-tape insert [156–160]. Other common means are to use stationary or fluid-driven rotating blades [161–165] or helically twisted pipes [166–167]. It has already been demonstrated in the literature [168] that swirl in a pipe flow has a long life, although there is no published numerical work that has dealt with an actual rotating fan blowing into the inlet of a pipe.

1.7.1   Physical Situation and Solution Strategy

A schematic rendering of the upstream portion of the physical situation is displayed in Figure 1.41. There, a round pipe is shown with a three-bladed axial fan mated

FIGURE 1.41 Schematic view of the upstream end of a round pipe with a rotating axial fan mated with its inlet cross section.

with its inlet cross section. The diagram clearly shows the physical supports of the stationary portion of the hub. The actual length of the pipe is drastically truncated in the view displayed in the figure. In actuality, two different pipe lengths are to be considered, one of which has a length-to-diameter ratio of 40 and the other a length-to-diameter ratio of 60.

Further insights into the physical situation can be seen in Figure 1.42a and b. These diagrams were prepared to illustrate enlargements of the solution domain. Figure 1.42a shows the radial enlargement of the domain to enable the fan to draw air laterally inward into its upstream face. Figure 1.42b displays the radial and axial enlargements to the side and upstream of the fan, respectively.

The fluid flow for this situation is turbulent, three-dimensional, and unsteady. Numerical solutions required discretization of the governing partial differential to the extent of approximately 40 million nodes and time steps as small as 0.001 s. The discretization and numerical solution were performed by means of ANSYS CFD 15.0 software, utilizing the SST model, with air as the flowing fluid. For the thermal problem, the pipe-wall boundary condition was uniform temperature. Two pipe lengths were investigated, having length/diameter ratios of 40 and 60, respectively.

In addition to the numerical solution based on fan rotation, two other solutions were obtained for purposes of comparison. One of these was based on the utilization of the fan/blower curve model. The manufacturer-supplied blower curve is displayed in Figure 1.43 along with system curves for the two considered pipe lengths. Note that, at the intersection points, the longer pipe displays a larger pressure drop and a lower volumetric flow rate as expected from intuition. The intersection points are the two operating points for which numerical solutions were performed.

For the numerical solutions, the blower-curve-determined volumetric flow rates were interpreted as steady and uniformly distributed across the pipe inlet section.

The other model used to obtain a second comparison with the rotating-fan-based predictions utilized the volumetric flow rate supplied by the actual rotating fan, and that flow rate was interpreted as steady and uniformly distributed at the pipe inlet.

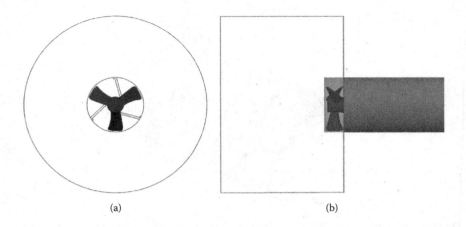

FIGURE 1.42 Diagrams displaying upstream extensions of the solution domain: (a) radial extension and (b) axial extension.

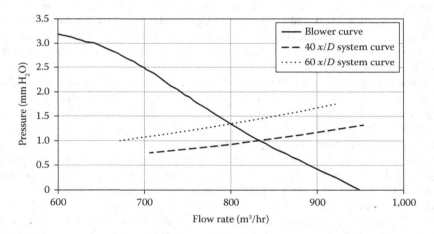

FIGURE 1.43 Blower curve for Sofasco D25089K 24V and system curves for pipes of length $x/D = 40$ and 60.

1.7.2 HEAT TRANSFER RESULTS AND DISCUSSION

Because the heat transfer varies both circumferentially and axially, there should—in principle—be two different heat transfer coefficients. However, in practice, it would be difficult to apply a circumferentially varying heat transfer coefficient. Instead, a circumferentially averaged heat transfer coefficient $\bar{h}$ is utilized and defined as

$$\bar{h} = \frac{1}{2\pi} \frac{\displaystyle\int_0^{2\pi} q(x, \theta)\, d\theta}{T_{\text{wall}} - T_{\text{bulk},x}} \tag{1.12}$$

in which $q(x, \theta)$ is the local heat flux, θ is the circumferential coordinate, x is the axial coordinate, T_{wall} is the uniform temperature of the pipe wall, and $T_{bulk,x}$ is the fluid bulk temperature at the axial location x. Corresponding to $\bar{h}$, an axially varying Nusselt number may be defined. It is

$$Nu_D = \frac{\bar{h}D}{k} \qquad (1.13)$$

The axial variations of the Nusselt number are plotted in Figures 1.44 and 1.45 for the pipes of length $L/D = 40$ and 60, respectively. In each figure, the Nusselt number is displayed as a function of x/D for the three considered cases. One of these is based on the rotating fan model, the second is based on the fan/blower curve model, and the third is based on a flow rate that is matched with that provided by the rotating fan. Note that, to achieve greater clarity at downstream locations, the main graph is supplemented by an inset at the upper right of each figure. Of the three curves displayed in each figure, only the rotating fan results are affected by swirl. The other two cases are actually conventional developing velocity and thermal pipe flows. In none of the three cases depicted in Figure 1.44 is thermal development achieved, whereas it appears that development is achieved in Figure 1.45.

The heat transfer results of Figure 1.44 are discussed first. It is shown that the swirl-related predictions (rotating fan) are much higher than the values for the comparison cases. For example, at $x/D = 5$, the Nusselt number for the fan-based model is about a factor of two larger than the value based on the blower curve model. If the same comparison is made at $x/D = 20$, the two predictions deviate by about 50%. The predictions based on the matched flow-rate model are lowest of all. This is because the flow rate borrowed from the rotating fan solution is lower than that taken from the blower curve and, additionally, there is no swirl in the case in question. Further inspection of Figure 1.44 reveals that, for the nonrotating fan cases, there appears to be an undershoot in the Nusselt number in the neighborhood of $x/D = 15$. This phenomenon has been observed previously [169–170] and was attributed to a transition from a laminar to a turbulent boundary layer. For a further perspective with respect to Figure 1.44, the Reynolds numbers for the three models displayed in the figure are 74,600 (blower curve model), 47,700 (rotating fan model), and 47,700 (matched flow-rate model).

Focus may now be directed to Figure 1.45. The general trends that have been identified in Figure 1.44 are repeated in Figure 1.45 but with some differences in detail. For one thing, the fully developed values of Nu_D for the blower curve and matched-flow rate models can be compared with those of the well-respected Petukhov–Popov formula, with the outcome being agreement to within 2% to 3%. It is also noteworthy that the rotating fan prediction at x/D of 60 is not quite fully developed, offering reinforcing testimony as to the long life of swirl.

Attention is now turned to local instantaneous heat transfer information. Figure 1.46 shows circumferential variations of the local wall heat flux at selected axial locations along the length of the $L/D = 40$ pipe. The figure shows that there is a consistent decrease in the magnitude of the heat flux with increasing

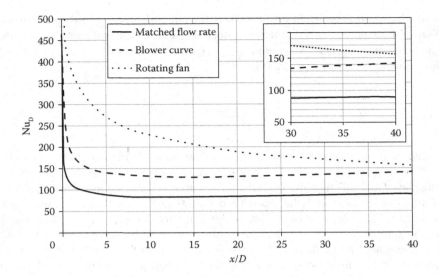

FIGURE 1.44 Axial variation of the circumferential-averaged Nusselt number as a function of x/D for a pipe length $L/D = 40$.

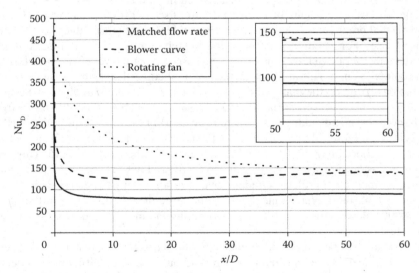

FIGURE 1.45 Axial variation of the circumferential-averaged Nusselt number as a function of x/D for a pipe length $L/D = 60$.

downstream distance. At any of the selected axial locations, it is seen that the heat flux undulates through a succession of maxima and minima. The number of such features appears to remain constant as the downstream distance increases.

The counterpart of Figure 1.46—but for the $L/D = 60$ pipe—is presented in Figure 1.47. The main trends of the earlier figure are more or less reproduced in

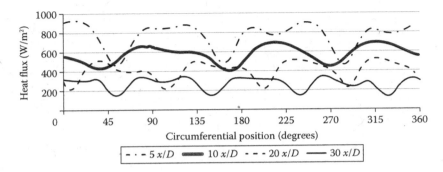

FIGURE 1.46 Circumferential variation of heat flux at specific axial locations for a pipe length of $L/D = 40$.

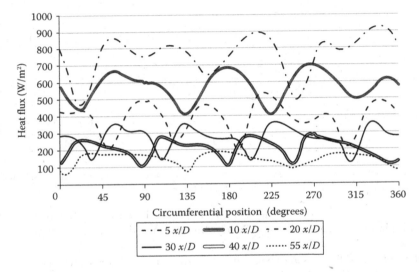

FIGURE 1.47 Circumferential variation of heat flux at specific axial locations for a pipe length of $L/D = 60$.

the current figure. Because the current figure is focused on a longer pipe, the magnitude of the heat flux reaches lower values than previously attained. The curves for the larger x/D locations tend to bunch up, indicating that the rate of heat flux decrease is diminishing.

1.7.3 Fluid Flow Results and Discussion

The presentation of the fluid flow results begins with vector diagrams that are presented in Figures 1.48 and 1.49 for normalized and unnormalized vectors, respectively. The vectors are displayed at three cross sections and lie in the plane of each cross section. The normalized vector diagrams of Figure 1.48 reveal an orderly rotational pattern that appears to be unchanging with increasing downstream distance.

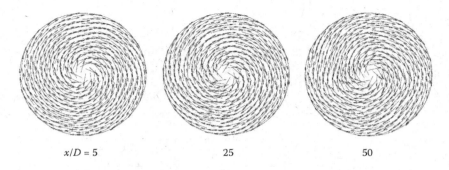

$x/D = 5$ 25   50

FIGURE 1.48   Normalized vector diagrams in cross-sectional planes that are situated at axial stations $x/D = 5$, 25 and at 50 for the $L/D = 60$ pipe.

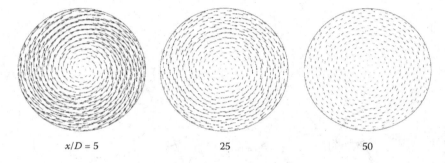

$x/D = 5$ 25   50

FIGURE 1.49   Unnormalized vector diagrams in cross-sectional planes that are situated at axial stations $x/D = 5$, 25 and at 50 for the $L/D = 60$ pipe.

The unnormalized vector diagrams shown in Figure 1.49 are more interesting. The magnitude of the rotating flow is clearly diminishing with increasing downstream distance.

The next fluid flow issue to be examined is the wall shear stress, results for which are displayed in Figure 1.50 for the $L/D = 60$ pipe. These results encompass the three models that have been explored throughout this section of the chapter: rotating fan model, blower-curve-based model, and matched flow-rate model. Not unexpectedly, the flow with swirl has a higher wall shear in the length of pipe in which the swirl is very vigorous. With decreasing swirl, the wall shear drops off markedly. The other two models that do not have swirl exhibit undershoots in their axial distributions in the neighborhood of $x/D = 15$. It is interesting and relevant to recall that the axial distributions of the circumferentially averaged Nusselt numbers displayed a similar behavior. It is believed that the cause of the local minima for these two processes, heat transfer and friction, are the same. That cause is hypothesized to be the transition of an initially laminar boundary layer into a turbulent boundary layer.

The shear stress is expected to level off to a fully developed value at sufficiently large downstream distances. Although it appears that the matched flow-rate case has more or less attained its fully developed value, the rotating fan case is continuing its development. When it does attain fully developed status, it may be expected that its

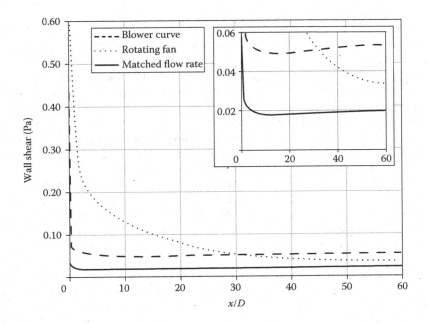

FIGURE 1.50 Axial variations of the circumferential-averaged wall shear stress as a function of x/D for a pipe length of $L/D = 60$.

shear stress value will be identical to that for the matched flow-rate case. This outcome can be attributed to the fact that both of these cases are characterized by the same Reynolds number. On the other hand, the fully developed value for the blower curve model must be higher than the others because its Reynolds number is also higher (74,600 versus 47,700).

1.7.4 RETROSPECTIVE VIEW OF THE INVESTIGATION OF ROTATING FAN FLOW DELIVERED TO A PIPE INLET

It was found that the swirl associated with a fan-delivered flow had a marked enhancing effect on the turbulent heat transfer in a round pipe. This conclusion is based on two comparisons. One of these is based on the flow model according to the fan/blower curve for the fan in question. A second comparison utilized the flow rate delivered by the rotating fan but modeled as a uniform inlet velocity without swirl. The swirl was seen to have a long life, exceeding 60 pipe diameters. In this regard, the present finding reinforces other information obtained in the past for other types of generated swirl.

For the swirl-free cases, both the circumferentially averaged Nusselt number and wall shear stress displayed an undershoot in their respective axial variations. The undershoot location was at approximately $x/D = 15$. This phenomenon is believed to be due to a transition of the wall-adjacent boundary layer from laminar to turbulent.

1.8 CONCLUDING REMARKS

The ongoing use of heat exchangers in many key industries provides strong motivation for the formulation and adoption of design methodologies that incorporate the most powerful and efficient tools. Steady improvement in the capacities and capabilities of computers has already greatly changed the design environment. Equally impacting on the design environment are the advances of software sophistication. These two closely linked events have enabled design to relate more closely to reality and to be more encompassing of interacting physical phenomena. In this light, the methodologies set forth in this chapter, although seemingly somewhat futuristic, are believed to be the right way to proceed and as soon as possible.

The most important message delivered by this chapter is that a specified heat transfer device cannot be regarded as having a unique performance in its own right. Rather, it has been demonstrated here that the actual performance depends on the interactions between the device and the fluid mover that serves it. A corollary of this message is the danger of designing a heat transfer device by basing the design on a manufacturer-supplied fan/blower curve. A blower curve provides a relationship between the magnitude of the delivered flow and the corresponding pressure rise imparted to the fluid flow. However, the magnitude of the fluid flow, although being of great importance to design, does not convey all the important features of the delivered fluid flow. Rather, other features such as swirl, eddies, backflow, cross-sectional nonuniformities, and unusually high turbulence may be embedded in the delivered flow.

In several of the applications treated here, swirl was shown to have a very important effect on heat transfer results. An important feature of the swirl's impact is its relatively long downstream life. The swirl creates a tornado-like tangential velocity distribution over the cross section such that the velocities are greatest in the neighborhood of the wall and least near the axis. This swirl distribution gives rise to a strong radial pressure variation, with the lowest pressures near the axis and the highest pressures near the wall. The movement of heat from the wall toward the axis of the pipe is conjectured to be enhanced by the pressure-driven radial inward flow.

REFERENCES

1. C.H. Huang, J.J. Lu, and H. Ay, A Three-Dimensional Heat Sink Module Design Problem with Experimental Verification, *International Journal of Heat and Mass Transfer*, vol. 54, pp. 1482–1492, 2011.
2. Z. Jian-Hui and Y. Chun-Xin, Design and Simulation of the CPU Fan and Heat Sinks, *IEEE Transactions on Components and Packaging Technologies*, vol. 31, pp. 890–903, 2008.
3. J. Stafford and F. Fortune, Investigation of Multiple Miniature Axial Fan Cooling Solutions and Thermal Modeling Approaches, *Journal of Electronic Packaging*, vol. 136, pp. 011008, 2014.
4. D. Nguyen, J.M. Gorman, E.M. Sparrow, and J.P. Abraham, Convective Heat Transfer Enhancement Versus Disenhancement: Impact of Fluid-Mover Characteristics, *Applied Thermal Engineering*, vol. 90, pp. 242–249, 2015.
5. V.P. Malapure, S.K. Mitra, and A. Bhattacharya, Numerical Investigation of Fluid Flow and Heat Transfer Over Louvered Fins in Compact Heat Exchanger, *International Journal of Thermal Sciences*, vol. 46, pp. 199–211, 2007.

6. M. Hiramatsu, T. Ishimaru, and K. Matsuzaki, Research on Fins For Air Conditioning Heat Exchangers: 1st Report, Numerical Analysis of Heat Transfer on Louvered Fins, *JSME International Journal, Series 2, Fluids Engineering, Heat Transfer, Power, Combustion, Thermophysical Properties*, vol. 33, pp. 749–756, 1990.

7. K. Suga and H. Aoki, Numerical Study on Heat Transfer and Pressure Drop in Multilouvered Fins, *Journal of Enhanced Heat Transfer*, vol. 2, pp. 231–238, 1995.

8. T. Perrotin and D. Clodic, Thermal-Hydraulic CFD Study in Louvered Fin-and-Flat-Tube Heat Exchangers, *International Journal of Refrigeration*, vol. 27, pp. 422–432, 2004.

9. K.N. Atkinsona, R. Drakulic, M.R. Heikal, and T.A. Cowell, Two-and Three-Dimensional Numerical Models of Flow and Heat Transfer Over Louvered Fin Arrays in Compact Heat Exchangers, *International Journal of Heat and Mass Transfer*, vol. 41, pp. 4063–4080, 1998.

10. Z. Čarija, B. Franković, M. Perčić, and M. Čavrak, Heat Transfer Analysis of Fin-and-Tube Heat Exchangers With Flat and Louvered Fin Geometries, *International Journal of Refrigeration*, vol. 45, pp. 160–167, 2014.

11. Z. Čarija and B. Franković, Heat Transfer Analysis of Flat and Louvered Fin-and-Tube Heat Exchangers Using CFD, *HEFAT2008, 6th International Conference on Heat Transfer, Fluid Mechanics and Thermodynamics*, pp. 1–6, 2008.

12. S. Stewart and S. Shelton, Design Study Comparison of Plain Finned Versus Louvered Finned-Tube Condenser Heat Exchangers, *ASME 2003 Heat Transfer Summer Conference*, pp. 703–710, 2003.

13. J.M. Gorman, M. Carideo, E.M. Sparrow, J.P. Abraham, Heat Transfer and Pressure Drop Comparison of Louver- and Plain-Finned Heat Exchangers Where One Fluid Passes through Flattened Tubes, *Case Studies in Thermal Engineering*, vol. 5, pp. 122–126, 2015.

14. H. Aoki, T. Shinagawa, and K. Suga, An Experimental Study of the Local Heat Transfer Characteristics in Automotive Louvered Fins, *Experimental Thermal and Fluid Science*, vol. 2, pp. 293–300, 1989.

15. Y.J. Chang and C.C. Wang, A Generalized Heat Transfer Correlation for Louver Fin Geometry, *International Journal of Heat and Mass Transfer*, vol. 40, pp. 533–544, 1997.

16. T.A. Cowell, M.R. Heikal, and A. Achaichia, Flow and Heat Transfer in Compact Louvered Fin Surfaces, *Experimental Thermal and Fluid Science*, vol. 10, pp. 192–199, 1995.

17. A. Sahnoun and R.L. Webb, Prediction of Heat Transfer and Friction for the Louver Fin Geometry, *Journal of Heat Transfer*, vol. 114, pp. 893–900, 1992.

18. M.H. Kim and C.W. Bullard, Air-Side Thermal Hydraulic Performance of Multi-Louvered Fin Aluminum Heat Exchangers, *International Journal of Refrigeration*, vol. 25, pp. 390–400, 2002.

19. C. Wang, K. Chen, J. Liaw, and C. Tseng, An Experimental Study of the Air-Side Performance of Fin-and-Tube Heat Exchangers Having Plain, Louver, and Semi-Dimple Vortex Generator Configuration, *International Journal of Heat and Mass Transfer*, vol. 80, pp. 281–287, 2015.

20. C. Wang, C. Lee, C. Chang, and Y. Chang, Some Aspects of Plate Fin-and-Tube Heat Exchangers: With and without Louvers, *Journal of Enhanced Heat Transfer*, vol. 6, pp. 357–368, 1999.

21. M.R. Shaeri and T.C. Jen, The Effects of Perforation Sizes on Laminar Heat Transfer Characteristics of an Array of Perforated Fins, *Energy Conversion and Management*, vol. 64, pp. 328–334, 2012.

22. M.F. Ismail, Effects of Perforations on the Thermal and Fluid Dynamic Performance of a Heat Exchanger, *IEEE Transactions on Components, Packaging and Manufacturing Technology*, vol. 3, pp. 1178–1185, 2013.

23. M.F. Ismail, M.N. Hasan, and M. Ali, Numerical Simulation of Turbulent Heat Transfer from Perforated Plate-Fin Heat Sinks, *Heat and Mass Transfer*, vol. 50, pp. 509–519, 2014.

24. E.M. Sparrow, B.R. Baliga, and S.V. Patankar, Forced Convection Heat Transfer from a Shrouded Fin Array with and without Tip Clearance, *Journal of Heat Transfer*, vol. 100, pp. 572–579, 1978.

25. E.M. Sparrow and S. Acharya, A Natural Convection Fin with a Solution-Determined Nonmonotonically Varying Heat Transfer Coefficient, *Journal of Heat Transfer*, vol. 103, pp. 218–225, 1981.

26. R.J. Moitsheki and A. Rowjee, Steady Heat Transfer Through a Two-Dimensional Rectangular Straight Fin, *Mathematical Problems in Engineering*, vol. 2011, p. 826819, 2011.

27. Y.M. Han, J.S. Cho, and H.S. Kang, Analysis of a One-Dimensional Fin Using the Analytic Method and the Finite Difference Method, *Korea Society For Industrial and Applied Mathematics*, vol. 9, pp. 91–98, 2005.

28. H.C. Ünal, Determination of the Temperature Distribution in an Extended Surface with a Nonuniform Heat Transfer Coefficient, *International Journal of Heat and Mass Transfer*, vol. 28, pp. 2279–2284, 1985.

29. D. Kim and S.J. Kim, Compact Modeling of Fluid Flow and Heat Transfer in Straight Fin Heat Sinks, *Journal of Electronic Packaging*, vol. 126, pp. 247–255, 2004.

30. S.W. Ma, A.I. Behbahani, and Y.G. Tsuei, Two-Dimensional Rectangular Fin with Variable Heat Transfer Coefficient, *International Journal of Heat and Mass Transfer*, vol. 34, pp. 79–85, 1991.

31. J.W. Yang, Periodic Heat Transfer in Straight Fins, *Journal of Heat Transfer*, vol. 94, pp. 310–314, 1972.

32. A.A. Joneidi, D.D. Ganji, and M. Babaelahi, Differential Transformation Method to Determine Fin Efficiency of Convective Straight Fins with Temperature Dependent Thermal Conductivity, *International Communications in Heat and Mass Transfer*, vol. 36, pp. 757–762, 2009.

33. G. Domairry and M. Fazeli, Homotopy Analysis Method to Determine the Fin Efficiency of Convective Straight Fins with Temperature-Dependent Thermal Conductivity, *Communications in Nonlinear Science and Numerical Simulation*, vol. 14, pp. 489–499, 2009.

34. C. Arslanturk, A Decomposition Method for Fin Efficiency of Convective Straight Fins with Temperature-Dependent Thermal Conductivity, *International Communications in Heat and Mass Transfer*, vol. 32, pp. 831–841, 2005.

35. M.H. Sharqawy and S.M. Zubair, Efficiency and Optimization of Straight Fins with Combined Heat and Mass Transfer–An Analytical Solution, *Applied Thermal Engineering*, vol. 28, pp. 2279–2288, 2008.

36. A.H. Elmahdy and R.C. Biggs, Efficiency of Extended Surfaces with Simultaneous Heat and Mass Transfer, *ASHRAE Transactions*, vol. 1, pp. 135–143, 1983.

37. C.J. Maday, The Minimum Weight One-Dimensional Straight Cooling Fin, *Journal of Manufacturing Science and Engineering*, vol. 96, pp. 161–165, 1974.

38. B. Kundu and P.K. Das, Performance and Optimization Analysis for Fins of Straight Taper with Simultaneous Heat and Mass Transfer, *Journal of Heat Transfer*, vol. 126, pp. 862–868, 2004.

39. F.S. Lai and Y.Y. Hsu Temperature Distribution in a Fin Partially Cooled by Nucleate Boiling, *AIChE Journal*, vol. 13, pp. 817–821, 1967.

40. R. Karvinen, Efficiency of Straight Fins Cooled by Natural or Forced Convection, *International Journal of Heat and Mass Transfer*, vol. 26, pp. 635–638, 1983.

41. H.T. Chen and J.C. Chou, Investigation of Natural-Convection Heat Transfer Coefficient on a Vertical Square Fin of Finned-Tube Heat Exchangers, *International Journal of Heat and Mass Transfer*, vol. 49, pp. 3034–3044, 2006.

42. A. Bar-Cohen, Fin Thickness for an Optimized Natural Convection Array of Rectangular Fins, *Journal of Heat Transfer*, vol. 101, pp. 564–566, 1979.

43. L.T. Yu, Application of Taylor Transformation to Optimize Rectangular Fins with Variable Thermal Parameters, *Applied Mathematical Modelling*, vol. 22, pp. 11–21, 1998.

44. K.T. Hong and R.L. Webb, Calculation of Fin Efficiency for Wet and Dry Fins, *HVAC&R Research*, vol. 2, pp. 27–41, 1996.

45. J.P. Luna-Abad and F. Alhama, Design and Optimization of Composite Rectangular Fins Using the Relative Inverse Thermal Admittance, *Journal of Heat Transfer*, vol. 135, p. 084504, 2013.

46. G.J. Huang and S.C. Wong, Dynamic Characteristics of Natural Convection from Horizontal Rectangular Fin Arrays, *Applied Thermal Engineering*, vol. 42, pp. 81–89, 2012.

47. M. Mehrtash and I. Tari, A Correlation for Natural Convection Heat Transfer from Inclined Plate-Finned Heat Sinks, *Applied Thermal Engineering*, vol. 51, pp. 1067–1075, 2013.

48. I.W. Chou and S.C. Wong, Natural Convection from Horizontal Rectangular Fin Arrays within Perforated Chassis, *Proceedings of the 2nd International Conference on Fluid Flow, Heat and Mass Transfer*, pp. 146-1–146-9, 2015.

49. H.J. Zhang and K.F. Sun, Conduction from Longitudinal Fin of Rectangular Profile with Exponential Vary Heat Transfer Coefficient, *Advanced Materials Research*, vol. 614, pp. 311–314, 2013.

50. S. Sharma and D. Prasad, A Comparative Analysis of Natural Convection between Horizontal and Vertical Heat Sink Using CFD. *International Journal of Engineering Research and Technology*, vol. 4, pp. 1089–1098, 2015.

51. P.L. Ndlovu and R.J. Moitsheki, Analytical Solutions for Steady Heat Transfer in Longitudinal Fins with Temperature-Dependent Properties, *Mathematical Problems in Engineering*, vol. 2013, p. 273052-1–273052-14, 2013.

52. P.L. Ndlovu and R.J. Moitsheki, Application of the Two-Dimensional Differential Transform Method to Heat Conduction Problem for Heat Transfer in Longitudinal Rectangular and Convex Parabolic Fins, *Communications in Nonlinear Science and Numerical Simulation*, vol. 18, pp. 2689–2698, 2013.

53. R.J. Moitsheki and C. Harley, Transient Heat Transfer in Longitudinal Fins of Various Profiles with Temperature-Dependent Thermal Conductivity and Heat Transfer Coefficient, *Pramana*, vol. 77, pp. 519–532, 2011.

54. R.J. Moitsheki, M.M. Rashidi, A. Basiriparsa, and A. Mortezaei, Analytical Solution and Numerical Simulation for One-Dimensional Steady Nonlinear Heat Conduction in a Longitudinal Radial Fin with Various Profiles, *Heat Transfer—Asian Research*, vol. 44, pp. 20–38, 2015.

55. M. Turkyilmazoglu, Heat Transfer through Longitudinal Fins, *Journal of Thermophysics and Heat Transfer*, vol. 28, pp. 806–811, 2014.

56. K.H. Dhanawade, V.K. Sunnapwar, and S.D. Hanamant, Thermal Analysis of Square and Circular Perforated Fin Arrays by Forced Convection, *International Journal of Cultural Engineering and Technology*, Special Issue-2, pp. 109–114, 2014.

57. S.A. El-Sayed, S.M. Mohamed, A.M. Abdel-latif, and E.A. Abdel-hamid, Investigation of Turbulent Heat Transfer and Fluid Flow in Longitudinal Rectangular-Fin Arrays of Different Geometries and Shrouded Fin Array, *Experimental Thermal and Fluid Science*, vol. 26, pp. 879–900, 2002.

58. F.E.M. Saboya, and E.M. Sparrow, Local and Average Transfer Coefficients for One-Row Plate Fin and Tube Heat Exchanger Configurations, *Journal of Heat Transfer*, vol. 96, pp. 265–272, 1974.

59. E.M. Sparrow and T.J. Beckey, Pressure Drop Characteristics for a Shrouded Longitudinal-Fin Array with Tip Clearance, *Journal of Heat Transfer*, vol. 103, pp. 393–395, 1981.

60. C.D. Jones and L.F. Smith, Optimum Arrangement of Rectangular Fins on Horizontal Surfaces for Free-Convection Heat Transfer, *Journal of Heat Transfer*, vol. 92, pp. 6–10, 1970.

61. H. Yüncü and G. Anbar, An Experimental Investigation on Performance of Rectangular Fins on a Horizontal Base in Free Convection Heat Transfer, *Heat and Mass Transfer*, vol. 33, pp. 507–514, 1998.

62. C.W. Leung, S.D. Probert, and M.J. Shilston, Heat Exchanger Design: Thermal Performances of Rectangular Fins Protruding from Vertical or Horizontal Rectangular Bases, *Applied Energy*, vol. 20, pp. 123–140, 1985.

63. C.W. Leung and S.D. Probert, Thermal Effectiveness of Short-Protrusion Rectangular, Heat-Exchanger Fins, *Applied Energy*, vol. 34, pp. 1–8, 1989.

64. C.W. Leung and S.D. Probert, Heat-Exchanger Performance: Effect of Orientation, *Applied Energy*, vol. 33, pp. 235–252, 1989.

65. Y.A. Cengel and T.H. Ngai, Cooling of Vertical Shrouded-Fin Arrays of Rectangular Profile by Natural Convection: An Experimental Study, *Heat Transfer Engineering*, vol. 12, pp. 27–39, 1991.

66. N.H. Saikhedkar and S.P. Sukhatme, Heat Transfer from Rectangular Cross-Sectioned Vertical Fin Arrays, *Proceedings of the Sixth National Heat and Mass Transfer Conference*, pp. 9–81, 1981.

67. S.G. Taji, G.V. Parishwad, and N.K. Sane, Enhanced Performance of Horizontal Rectangular Fin Array Heat Sink Using Assisting Mode of Mixed Convection, *International Journal of Heat and Mass Transfer*, vol. 72, pp. 250–259, 2014.

68. S.G. Taji, G.V. Parishwad, and N.K. Sane, Experimental Investigation of Heat Transfer and Flow Pattern from Heated Horizontal Rectangular Fin Array Under Natural Convection, *Heat and Mass Transfer*, vol. 50, pp. 1005–1015, 2014.

69. M. Dogan and M. Sivrioglu, Experimental Investigation of Mixed Convection Heat Transfer from Longitudinal Fins in a Horizontal Rectangular Channel: In Natural Convection Dominated Flow Regimes, *Energy Conversion and Management*, vol. 50, pp. 2513–2521, 2009.

70. M.R. Shaeri, M. Yaghoubi, and K. Jafarpur, Heat Transfer Analysis of Lateral Perforated Fin Heat Sinks, *Applied Energy*, vol. 86, pp. 2019–2029, 2009.

71. S.H. Yu and K.S. Lee, Heat Transfer from Rectangular Fins with a Circular Base, *Transactions of the Korean Society of Mechanical Engineers B*, vol. 35, pp. 467–472, 2011.

72. S.D. Suryawanshi and N.K. Sane, Natural Convection Heat Transfer from Horizontal Rectangular Inverted Notched Fin Arrays, *Journal of Heat Transfer*, vol. 131, p. 082501, 2009.

73. U. Engdar and J. Klingmann, Investigation of Two-Equation Turbulence Models Applied to a Confined Axis-Symmetric Swirling Flow. *ASME 2002 Pressure Vessels and Piping Conference*, pp. 199–206, 2002.

74. E.M. Sparrow, J.M. Gorman, and J.P. Abraham, Quantitative Assessment of the Overall Heat Transfer Coefficient U, *ASME Journal of Heat Transfer*, vol. 135, p. 061102, 2013.

75. Y. Bayazit, E.M. Sparrow, and D.D. Joseph, Perforated Plates for Fluid Management: Plate Geometry Effects and Flow Regimes, *International Journal of Thermal Sciences*, vol. 85, pp. 104–111, 2014.

76. A. Li, X. Chen, and L. Chen, Numerical Investigations on Effects of Seven Drag Reduction Components in Elbow and T-Junction Close-Coupled Pipes, *Building Services Engineering Research and Technology*, vol. 36, pp. 295–310, 2015.

77. S.W. Churchill, A Reinterpretation of the Turbulent Prandtl Number, *Industrial & Engineering Chemistry Research*, vol. 41, pp. 6393–6401, 2002.

78. W.M. Kays, Turbulent Prandtl Number—Where Are We?, *ASME Journal Heat Transfer*, vol. 116, pp. 284–295, 1994.
79. T.M. Hsieh, C.H. Luo, F.C. Huang, J.H. Wang, L.J. Chien, and G.B. Lee, Enhancement of Thermal Uniformity for a Microthermal Cycler and its Application for Polymerase Chain Reaction, *Sensors and Actuators B: Chemical*, vol. 130, pp. 848–856, 2008.
80. V. Raghavan, S.E. Whitney, R.J. Ebmeier, N.V. Padhye, M. Nelson, H.J. Viljoen, and G. Gogos, Thermal Analysis of the Vortex Tube Based Thermocycler for Fast DNA Amplification: Experimental and Two-Dimensional Numerical Results, *Review of Scientific Instruments*, vol. 77, p. 094301, 2006.
81. D.S. Lee, C.Y. Tsai, W.H. Yuan, P.J. Chen, and P.H. Chen, A New Thermal Cycling Mechanism for Effective Polymerase Chain Reaction in Microliter Volumes, *Microsystem Technologies*, vol. 10, pp. 579–584, 2004.
82. D.S. Yoon, Y.-S. Lee, Y. Lee, H.J. Cho, S.W. Sung, K.W. Oh, J. Cha, and G. Lim, Precise Temperature Control and Rapid Thermal Cycling in a Micromachined DNA Polymerase Chain Reaction Chip, *Journal of Micromechanics and Microengineering*, vol. 12, pp. 813–823, 2002.
83. J. Chiou, P. Matsudaira, A. Sonin, and D. Ehrlich, A Closed-Cycle Capillary Polymerase Chain Reaction Machine, *Analytical Chemistry*, vol. 73, pp. 2018–2021, 2001.
84. C.J. Kobus and T. Oshio, Development of a Theoretical Model for Predicting the Thermal Performance Characteristics of a Vertical Pin-Fin Array Heat Sink Under Combined Forced and Natural Convection with Impinging Flow, *International Journal of Heat and Mass Transfer*, vol. 48, pp. 1053–1063, 2005.
85. J.G. Maveety and H.H. Jung, Design of an Optimal Pin-Fin Heat Sink with Air Impingement Cooling, *International Communications in Heat and Mass Transfer*, vol. 27, pp. 229–240, 2000.
86. Y.T. Yang and H.S. Peng, Numerical Study of Pin-Fin Heat Sink with Un-Uniform Fin Height Design, *International Journal of Heat and Mass Transfer*, vol. 51, pp. 4788–4796, 2008.
87. H.I. You and C.H. Chang, Numerical Prediction of Heat Transfer Coefficient for a Pin-Fin Channel Flow, *Journal of Heat Transfer*, vol. 119, pp. 840–843, 2007.
88. R. Bahadur and A. Bar-Cohen, Orthotropic Thermal Conductivity Effect on Cylindrical Pin Fin Heat Transfer, *International Journal of Heat and Mass Transfer*, vol. 50, pp. 1155–1162, 2007.
89. E. Shaukatullah, W.R. Storr, B.J. Hansen, and M. Gaynes, Design and Optimization of Pin Fin Heat Sinks For Low Velocity Applications, *Proceedings Twelfth Annual IEEE Semiconductor Thermal Measurement and Management Symposium, SEMI-THERM XII*, pp. 151–163, 1996.
90. Y.C. Yang, H.L. Lee, E.J. Wei, J.F. Lee, and T.S. Wu, Numerical Analysis of Two Dimensional Pin Fins with Non-Constant Base Heat Flux, *Energy Conversion and Management*, vol. 46, pp. 881–892, 2005.
91. S.S. Chu and W.J. Chang, Hybrid Numerical Method for Transient Analysis of Two-Dimensional Pin Fins with Variable Heat Transfer Coefficients, *International Communications in Heat and Mass Transfer*, vol. 29, pp. 367–376, 2002.
92. R.S.R. Gorla and I. Pop, Conjugate Heat Transfer with Radiation from a Vertical Circular Pin in a Non-Newtonian Ambient Medium, *Wärme-und Stoffübertragung*, vol. 28, pp. 11–15, 1993.
93. U.C. Chen, W.J. Chang, and J.C. Hsu, Two-Dimensional Inverse Problem in Estimating Heat Flux of Pin Fins, *International Communications in Heat and Mass Transfer*, vol. 28, pp. 793–801, 2001.
94. D. Poulikakos and A. Bejan, Fin Geometry For Minimum Entropy Generation in Forced Convection, *Journal of Heat Transfer*, vol. 104, pp. 616–623, 1982.

95. P. Malekzadeh and H. Rahideh, IDQ Two-Dimensional Nonlinear Transient Heat Transfer Analysis of Variable Section Annular Fins, *Energy Conversion and Management*, vol. 48, pp. 269–276, 2007.

96. H.S. Kang, Optimization of a Pin Fin with Variable Base Thickness, *Journal of Heat Transfer*, vol. 132, p. 034501, 2010.

97. L. Zhang and S.L. Liu, Numerical Study on Heat Transfer and Pressure Drop in Pin-Fin Channels with Lateral Flow Ejection, *Journal of Propulsion Technology Beijing*, vol. 25, pp. 307–310, 2004.

98. W.J. Chang, U.C. Chen, and H.M. Chou, Transient Analysis of Two-Dimensional Pin Fins with Non-Constant Base Temperature, *JSME International Journal Series B*, vol. 45, pp. 331–337, 2002.

99. F. Hajabdollahi, H.H. Rafsanjani, Z. Hajabdollahi, and Y. Hamidi, Multi-Objective Optimization of Pin Fin to Determine the Optimal Fin Geometry Using Genetic Algorithm, *Applied Mathematical Modelling*, vol. 36, pp. 244–254, 2012.

100. S. Kanyakam and S. Bureerat, Multiobjective Optimization of a Pin-Fin Heat Sink Using Evolutionary Algorithms, *Journal of Electronic Packaging*, vol. 134, p. 021008, 2012.

101. S. Pashah, A.F.M. Arif, and S.M. Zubair, Study of Orthotropic Pin Fin Performance Through Axisymmetric Thermal Non-Dimensional Finite Element, *Applied Thermal Engineering*, vol. 31, pp. 376–384, 2011.

102. L.C.M.T. Li-Ting and Z.D.M.G. Zhi-Qiang, Investigation of the Best Augmented Heat Transfer of Pin-Fin Arrays in Passage For Trailing Edge of Turbine Blade, *Journal of Engineering Thermophysics*, vol. 4, p. 029, 2006.

103. L.I. Díez, A. Campo, and C. Cortés, Quick Design of Truncated Pin Fins of Hyperbolic Profile For Heat-Sink Applications by Using Shortened Power Series, *Applied Thermal Engineering*, vol. 29, pp. 815–821, 2009.

104. K. Park, P.K. Oh, and H.J. Lim, Optimum Design of a Pin-Fins Type Heat Sink Using the CFD and Mathematical Optimization, *International Journal of Air-Conditioning and Refrigeration*, vol. 13, pp. 71–82, 2005.

105. P. Razelos, The Optimum Dimensions of Convective Pin Fins with Internal Heat Generation, *Journal of the Franklin Institute*, vol. 321, pp. 1–19, 1986.

106. A. Radmehr, K.M. Kelkar, P. Kelly, S.V. Patankar, and S.S. Kang, Analysis of the Effect of Bypass on the Performance of Heat Sinks Using Flow Network Modeling (FNM), *Fifteenth Annual IEEE Semiconductor Thermal Measurement and Management Symposium*, pp. 42–47, 1999.

107. C. Ling, C. Min, and Y. Xiao, Three Dimensional Numerical Investigation on the Effect of Wedge Angle of a Passage with Pin-Fin Arrays, *Journal of Thermal Science*, vol. 13, pp. 138–142, 2004.

108. H.S. Kang, Optimization of a Pin Fin Based on the Increasing Rate of Heat Loss, *Journal of the Korean Society for Industrial and Applied Mathematics*, vol. 12, pp. 25–32, 2008.

109. B. Kundu, and K.S. Lee, Shape Optimization for the Minimum Volume of Pin Fins in Simultaneous Heat and Mass Transfer Environments, *Heat and Mass Transfer*, vol. 48, pp. 1333–1343, 2012.

110. W.A. Khan, J.R. Culham, and M.M. Yovanovich, Modeling of Cylindrical Pin-Fin Heat Sinks for Electronic Packaging, *IEEE Transactions on Components and Packaging Technologies*, vol. 31, pp. 536–545, 2008.

111. S.B. Chin, J.J. Foo, Y.L. Lai, and T.K.K. Yong, Forced Convective Heat Transfer Enhancement with Perforated Pin Fins, *Heat and Mass Transfer*, vol. 49, pp. 1447–1458, 2013.

112. A.K. Saha and S. Acharya, Parametric Study of Unsteady Flow and Heat Transfer in a Pin-Fin Heat Exchanger, *International Journal of Heat and Mass Transfer*, vol. 46, pp. 3815–3830, 2003.

113. D. Kim, S.J. Kim, and A. Ortega, Compact Modeling of Fluid Flow and Heat Transfer in Pin Fin Heat Sinks, *Journal of Electronic Packaging*, vol. 126, pp. 342–350, 2004.

114. P.A. Deshmukh and R.M. Warkhedkar, Thermal Performance of Elliptical Pin Fin Heat Sink Under Combined Natural and Forced Convection, *Experimental Thermal and Fluid Science*, vol. 50, pp. 61–68, 2013.

115. W. Duangthongsuk and S. Wongwises, A Comparison of the Heat Transfer Performance and Pressure Drop of Nanofluid-Cooled Heat Sinks with Different Miniature Pin Fin Configurations, *Experimental Thermal and Fluid Science*, vol. 69, pp. 111–118, 2015.

116. E.M. Sparrow, J.W. Ramsey, and C.A.C. Altemani, Experiments on In-Line Pin Fin Arrays and Performance Comparisons with Staggered Arrays, *Journal of Heat Transfer*, vol. 102, pp. 44–50, 1980.

117. E.M. Sparrow and J.W. Ramsey, Heat Transfer and Pressure Drop for a Staggered Wall-Attached Array of Cylinders with Tip Clearance, *International Journal of Heat and Mass Transfer*, vol. 21, pp. 1369–1378, 1978.

118. E.M. Sparrow and E.D. Larson, Heat Transfer From Pin-Fins Situated in an Oncoming Longitudinal Flow Which Turns to Crossflow, *International Journal of Heat and Mass Transfer*, vol. 25, pp. 603–614, 1982.

119. B.A. Brigham and G.J. Vanfossen, Length to Diameter Ratio and Row Number Effects in Short Pin Fin Heat Transfer, *ASME Transactions, Journal of Engineering for Gas Turbines and Power*, vol. 106, pp. 241–245, 1984.

120. R.J. Goldstein, M.Y. Jabbari, and S.B. Chen, Convective Mass Transfer and Pressure Loss Characteristics of Staggered Short Pin-Fin Arrays, *International Journal of Heat and Mass Transfer*, vol. 37, pp. 149–160, 1994.

121. S.C. Lau, J.C. Han, and Y.S. Kim, Turbulent Heat Transfer and Friction in Pin Fin Channels with Lateral Flow Ejection, *Journal of Heat Transfer*, vol. 111, pp. 51–58, 1989.

122. J.J. Wei and H. Honda, Effects of Fin Geometry on Boiling Heat Transfer from Silicon Chips with Micro-Pin-Fins Immersed in FC-72, *International Journal of Heat and Mass Transfer*, vol. 46, pp. 4059–4070, 2003.

123. Z. Chen, Q. Li, D. Meier, and H.J. Warnecke, Convective Heat Transfer and Pressure Loss in Rectangular Ducts with Drop-Shaped Pin Fins, *Heat and Mass Transfer*, vol. 33, pp. 219–224, 1997.

124. M. Tahat, Z.H. Kodah, B.A. Jarrah, and S.D. Probert, Heat Transfers from Pin-Fin Arrays Experiencing Forced Convection, *Applied Energy*, vol. 67, pp. 419–442, 2000.

125. A. Koşar and Y. Peles, Boiling Heat Transfer in a Hydrofoil-Based Micro Pin Fin Heat Sink, *International Journal of Heat and Mass Transfer*, vol. 50, pp. 1018–1034, 2007.

126. S.C. Lau, J.C. Han, and T. Batten, Heat Transfer, Pressure Drop, and Mass Flow Rate in Pin Fin Channels with Long and Short Trailing Edge Ejection Holes, *Journal of Turbomachinery*, vol. 111, pp. 116–123, 1989.

127. L. Brignoni and S.V. Garimella, Experimental Optimization of Confined Air Jet Impingement on a Pin Fin Heat Sink, *IEEE Transactions on Components and Packaging Technologies*, vol. 22, pp. 399–404, 1999.

128. J.J. Hwang, D.Y. Lai, and Y.P. Tsia, Heat Transfer and Pressure Drop in Pin-Fin Trapezoidal Ducts, *Journal of Turbomachinery*, vol. 121, pp. 264–271, 1999.

129. K. Al-Jamal and H. Khashashneh, Experimental Investigation in Heat Transfer of Triangular and Pin Fin Arrays, *Heat and Mass Transfer*, vol. 34, pp. 159–162, 1998.

130. R.T. Huang, W.J. Sheu, and C.C. Wang, Orientation Effect on Natural Convective Performance of Square Pin Fin Heat Sinks, *International Journal of Heat and Mass Transfer*, vol. 51, pp. 2368–2376, 2008.

131. R. Matsumoto, S. Kikkawa, and M. Senda, Effect of Pin Fin Arrangement on Endwall Heat Transfer, *JSME International Journal Series B*, vol. 40, pp. 142–151, 1997.

132. M.K. Chyu, S.C. Siw, and H.K. Moon, Effects of Height-to-Diameter Ratio of Pin
     Element on Heat Transfer from Staggered Pin-Fin Arrays, *ASME Turbo Expo 2009:
     Power for Land, Sea, and Air*, pp. 705–713, 2009.
133. N. Sahiti, F. Durst, and A. Dewan, Heat Transfer Enhancement by Pin Elements,
     International Journal of Heat and Mass Transfer, vol. 48, pp. 4738–4747, 2005.
134. E.A.M. Elshafei, Natural Convection Heat Transfer from a Heat Sink with Hollow/
     Perforated Circular Pin Fins, *Energy*, vol. 35, pp. 2870–2877, 2010.
135. M.K. Chyu, E.O. Oluyede, and H.K. Moon, Heat Transfer on Convective Surfaces with
     Pin-Fins Mounted in Inclined Angles, *ASME Turbo Expo 2007: Power for Land, Sea,
     and Air*, pp. 861–869, 2007.
136. S.A. Lawson, A.A. Thrift, K.A. Thole, and A. Kohli, Heat Transfer From Multiple
     Row Arrays of Low Aspect Ratio Pin Fins, *International Journal of Heat and Mass
     Transfer*, vol. 54, pp. 4099–4109, 2011.
137. L.M. Wright, E. Lee, and J.C. Han, Effect of Rotation on Heat Transfer in Rectangular
     Channels with Pin-Fins, *Journal of Thermophysics and Heat Transfer*, vol. 18,
     pp. 263–272, 2004.
138. J.S. Park, K.M. Kim, D.H. Lee, H.H. Cho, and M.K. Chyu, Heat Transfer on Rotating
     Channel with Various Heights of Pin-Fin, *ASME Turbo Expo 2008: Power for Land,
     Sea, and Air*, pp. 727–734, 2008.
139. H.C. Ryu, D. Kim, and S.J. Kim, Experimental Analysis of Shrouded Pin Fin Heat
     Sinks for Electronic Equipment Cooling, *The Eighth Intersociety Conference on
     Thermal and Thermomechanical Phenomena in Electronic Systems, IEEE ITHERM*,
     pp. 261–266, 2002.
140. P.S. Lee, J.C. Ho, and H. Xue, Experimental Study on Laminar Heat Transfer in
     Microchannel Heat Sink, *The Eighth Intersociety Conference on Thermal and
     Thermomechanical Phenomena in Electronic Systems, IEEE ITHERM 2002*,
     pp. 379–386, 2002.
141. Y. Rao, C. Wan, and S. Zang, Comparisons of Flow Friction and Heat Transfer
     Performance in Rectangular Channels with Pin Fin-Dimple, Pin Fin and Dimple
     Arrays, *ASME Turbo Expo 2010: Power for Land, Sea, and Air*, pp. 185–195, 2010.
142. S. Dong, S. Liu, and H. Su, An Experimental Investigation of Heat Transfer In Pin Fin
     Array, *Heat Transfer—Asian Research*, vol. 30, pp. 533–541, 2001.
143. R. Babyand and C. Balaji, Thermal Management of Electronics Using Phase Change
     Material Based Pin Fin Heat Sinks, *Journal of Physics: Conference Series*, vol. 395,
     p. 012134, 2012.
144. A.B. Dhumne and H.S. Farkade, Heat Transfer Analysis of Cylindrical Perforated
     Fins in Staggered Arrangement, *International Journal of Engineering Science and
     Technology*, vol. 6, pp. 125–138, 2013.
145. Q. Li, L. Ma, Z. Chen, and H.J. Warnecke, Heat Transfer Characteristics of a Tube with
     Elliptic Pin Fins in Crossflow of Air, *Heat and Mass Transfer*, vol. 39, pp. 529–533,
     2003.
146. H. El-Sheikh and S.V. Gurimella, Enhancement of Air Jet Impingement Heat Transfer
     Using Pin-Fin Heat Sinks, *IEEE Transactions on Components and Packaging
     Technologies*, vol. 23, pp. 300–308, 2000.
147. J.W. Baughn, P.T. Ireland, T.V. Jones, and N. Saniei, A Comparison of the Transient and
     Heated-Coating Methods For the Measurement of Local Heat Transfer Coefficients
     on a Pin Fin, *ASME 1988 International Gas Turbine and Aeroengine Congress*,
     p. V004T09A031, 1998.
148. T. Aihara, S. Maruyama, and S. Kobayakawa, Free Convective/Radiative Heat
     Transfer from Pin-Fin Arrays with a Vertical Base Plate (General Representation of
     Heat Transfer Performance), *International Journal of Heat and Mass Transfer*, vol. 33,
     pp. 1223–1232, 1990.

149. G.J. VanFossen, Heat-Transfer Coefficients for Staggered Arrays of Short Pin Fins, *Journal of Engineering for Gas Turbines and Power*, vol. 104, pp. 268–274, 1982.
150. E.M. Sparrow and S.B. Vemuri, Orientation Effects on Natural Convection/Radiation Heat Transfer from Pin-Fin Arrays, *International Journal of Heat and Mass Transfer*, vol. 29, pp. 359–368, 1986.
151. D.E. Metzger, W.B. Shepard, and S.W. Haley, Row Resolved Heat Transfer Variations in Pin-Fin Arrays Including Effects of Nonuniform Arrays and Flow Convergence, *ASME 1986 International Gas Turbine Conference and Exhibit*, p. V004T09A015, 1986.
152. S.J. Park, D. Jang, S.J. Yook, and K.S. Lee, Optimization of a Staggered Pin-Fin for a Radial Heat Sink Under Free Convection, *International Journal of Heat and Mass Transfer*, vol. 87, pp. 184–188, 2015.
153. E.M. Sparrow and V.B. Grannis, Pressure Drop Characteristics of Heat Exchangers Consisting of Arrays of Diamond-Shaped Pin Fins, *International Journal of Heat and Mass Transfer*, vol. 34, pp. 589–600, 1991.
154. E. Yu and Y. Joshi, Heat Transfer Enhancement From Enclosed Discrete Components Using Pin–Fin Heat Sinks, *International Journal of Heat and Mass Transfer*, vol. 45, pp. 4957–4966, 2002.
155. M.B. Dogruoz, M. Urdaneta, and A. Ortega, Experiments and Modeling of the Hydraulic Resistance and Heat Transfer of In-Line Square Pin Fin Heat Sinks with Top By-Pass Flow, *International Journal of Heat and Mass Transfer*, vol. 48, pp. 5058–5071, 2005.
156. S.K. Agarwal and M. Raja Rao, Heat Transfer Augmentation for the Flow of a Viscous Liquid in Circular Tubes Using Twisted Tape Inserts, *International Journal of Heat and Mass Transfer*, vol. 39, pp. 3547–3557, 1996.
157. R.M. Manglik and A.E. Bergles, Swirl Flow Heat Transfer and Pressure Drop with Twisted-Tape Inserts, *Advances in Heat Transfer*, vol. 36, pp. 183–266, 2003.
158. S. Eiamsa-Ard, K. Wongcharee, and S. Sripattanapipat, 3-D Numerical Simulation of Swirling Flow and Convective Heat Transfer in a Circular Tube Induced by Means of Loose-Fit Twisted Tapes, *International Communications in Heat and Mass Transfer*, vol. 36, pp. 947–955, 2009.
159. R. Beigzadeh, M. Rahimi, M. Parvizi, and S. Eiamsa-ard, Application of ANN and GA for the Prediction and Optimization of Thermal and Flow Characteristics in a Rectangular Channel Fitted with Twisted Tape Vortex Generators, *Numerical Heat Transfer, Part A: Applications*, vol. 65, pp. 186–199, 2014.
160. E.Y. Rios-Iribe, M.E. Cervantes-Gaxiola, E. Rubio-Castro, J.M. Ponce-Ortega, M.D. González-Llanes, C. Reyes-Moreno, and O.M. Hernández-Calderón, Heat Transfer Analysis of a Non-Newtonian Fluid Flowing Through a Circular Tube with Twisted Tape Inserts, *Applied Thermal Engineering*, vol. 84, pp. 225–236, 2015.
161. B.A. Saraç and T. Bali, An Experimental Study on Heat Transfer and Pressure Drop Characteristics of Decaying Swirl Flow Through a Circular Pipe with a Vortex Generator, *Experimental Thermal and Fluid Science*, vol. 32, pp. 158–165, 2007.
162. M. Ahmadvand, A.F. Najafi, and S. Shahidinejad, An Experimental Study and CFD Analysis Towards Heat Transfer and Fluid Flow Characteristics of Decaying Swirl Pipe Flow Generated by Axial Vanes, *Meccanica*, vol. 45, pp. 111–129, 2010.
163. A.E. Zohir and A.G. Gomaa, Heat Transfer Enhancement Through Sudden Expansion Pipe Airflow Using Swirl Generator with Different Angles, *Experimental Thermal and Fluid Science*, vol. 45, pp. 146–154, 2013.
164. W. Duangthongsuk and S. Wongwises, An Experimental Investigation of The Heat Transfer and Pressure Drop Characteristics of a Circular Tube Fitted With Rotating Turbine-Type Swirl Generators, *Experimental Thermal and Fluid Science*, vol. 45, pp. 8–15, 2013.

165. H. Seo, S.D. Park, S.B. Seo, H. Heo, and I.C. Bang, Swirling Performance of Flow-Driven Rotating Mixing Vane toward Critical Heat Flux Enhancement, *International Journal of Heat and Mass Transfer*, vol. 89, pp. 1216–1229, 2015.

166. J.M. Gorman, E.M. Sparrow, J.P. Abraham, and G.S. Mowry, Operating Characteristics and Fabrication of a Uniquely Compact Helical Heat Exchanger, *Applied Thermal Engineering*, vol. 5, pp. 1070–1075, 2012.

167. J.M. Gorman, E.M. Sparrow, G.S. Mowry, and J.P. Abraham, Simulation of Helically Wrapped, Compact Heat Exchangers, *Journal of Renewable and Sustainable Energy*, vol. 3, p. 043120, 2011.

168. E.M. Sparrow and A. Chaboki, Swirl-Affected Turbulent Fluid Flow and Heat Transfer in a Circular Tube, *Journal of Heat Transfer*, vol. 106, pp. 766–773, 1984.

169. A.F. Mills, Experimental Investigation of Turbulent Heat Transfer in the Entrance Region of a Circular Conduit, *Journal of Mechanical Engineering Science*, vol. 4, pp. 63–77, 1962.

170. F. Kreith and M. Bohn, *Principles of Heat Transfer* (6th ed.), Brooks/Cole, Boston, MA, 2001.

2 On Computational Heat Transfer Procedures for Heat Exchangers in Single-Phase Flow Operation

Bengt Sundén

CONTENTS

ABSTRACT: This chapter presents a brief summary and state-of-the-art overview of computational methods in heat transfer equipment, that is, computational fluid dynamics (CFD) methods for thermal problems and turbulence modeling for single-phase applications in the design, research, and development of heat exchangers. Some examples for real heat exchangers are shown to demonstrate how CFD methods can be used as research and development tools. Nevertheless, limitations and shortcomings as well as the need for further research are highlighted. Also specific methods such as artificial neural networks and genetic algorithms are described, and examples are given as to how these can be used.

2.1 INTRODUCTION

In the design and development processes of heat exchangers, one needs to calculate or estimate the performance of the heat exchangers, and different methods are possible for this. One method is to adopt engineering approaches, which are generally described in textbooks or handbooks on heat transfer and heat exchangers [1–5], where rating and sizing issues are discussed. Another method suitable for research and development is to adopt computational fluid dynamics (CFD) approaches [6,7], which also allow a detailed analysis of basic transport phenomena within heat exchangers. CFD is especially useful in the initial design steps by reducing the number of prototypes and providing a good insight in the transport processes occurring in the heat exchanger. Also, heat exchanger optimization is important for process intensification, and CFD might be a helpful tool here as well. However, for optimization, methods based on artificial neural networks (ANNs) and genetic algorithms (GAs) might be used [8,9].

The computational modeling approaches have been developed extensively and now allow the performance of numerical simulations of complex problems and the investigation of the influence of various design parameters. Detailed distributions of the relevant variables can be obtained, and insights into the underlying physical processes can be gained.

Computers and computational techniques currently available enable very detailed representation of actual physical and engineering problems, and a comprehensive set of models is available for the relevant physical processes. Massive parallel computations are now possible as well.

There are several reasons why heat transfer and heat exchangers play key roles in the reduction of greenhouse gas emissions and for achieving sustainable development [10,11]. New advanced techniques reducing energy consumption and improving power conversion efficiencies, introduction of fuel cells, usage of renewable energy sources, hydrogen production, application of exhaust gas recirculation and cooling, and so forth all call for heat transfer analysis and introduction of heat exchangers.

In most heat exchangers, the heat transfer is dominated by convection and conduction from a hot to a cold fluid separated by solid walls. However, for high-temperature applications, thermal radiation might be of significance as well [12]. At low temperatures (cryogenic state), multistream heat exchangers are frequently utilized [13]. Design and sizing of heat exchangers involves many complex procedures [14]. The total amount of heat transferred, pressure drops, performance efficiency, and the manufacturing and operating costs are important in the final design. In some cases, the overall cost is important, although in other applications, weight and size are the most vital factors.

As a heat exchanger is designed, the convective heat transfer coefficients between fluids and walls are important. Commonly, the hot and cold fluids flow in ducts of various designs. Thus, internal duct flow is very important in the field of heat exchangers. These heat transfer coefficients are dependent on the flow velocity, fluid properties, duct cross section geometry and size, and on duct length. For simple geometries, such coefficients are available in the literature [15,16] on single-phase flow at laminar, turbulent, or transitional conditions. In the development of compact heat exchangers (CHEs), it is important to decrease the heat transfer surface area. However, if the same or a higher amount of heat flow rate prevails, the only way out is to improve the heat transfer coefficients and thus reduce the thermal resistances on the hot and cold sides. Thus, the concept of enhanced heat transfer [17] is very relevant for research and development of efficient heat exchangers. New surfaces and surface modifications such as cross-corrugated surfaces, dimples, ribbed surfaces, offset strip fins, louvered surfaces, inserts (turbulators), wavy ducts, ducts with bumps, surfaces with winglets, and so forth, are being developed extensively.

Attempts to provide efficient, compact, and inexpensive heat exchangers are indeed challenging. To achieve this, both theoretical and experimental investigations must be conducted, and advanced modern methods must be adopted. The present chapter focuses on modeling approaches and how they can be used in the research, design, and development of heat exchangers. CFD is a numerical solution methodology of governing equations for mass conservation, momentum, and heat transfer and other transport processes. When the focus is on thermal issues, computational heat transfer (CHT) and numerical heat transfer (NHT) are more appropriate names.

Shah et al. [18] presented a summary of literature on numerical analysis of CHE surfaces. The pioneering work on shell-and-tube heat exchangers was presented by Patankar and Spalding [19]. The idea here was to consider the shell side as a porous medium and to treat the presence of tubes and baffles by introducing volume porosities and surface permeability. In addition, flow and thermal resistances also have to be introduced. Later follow-up studies include those by Prithiviraj and Andrews [20,21] and He et al. [22].

This chapter presents current CFD methods for single-phase flows, including turbulence modeling and associated problems and limitations, as well as providing examples of CFD applications in a variety of heat exchanger problems. Some brief reviews have been presented earlier by Sunden [23] and Bhutta et al. [24]. This chapter also presents results from applications of commercially available computer codes and in-house codes. In addition, the application of ANNs and GAs to heat exchanger tasks are presented.

2.2 GOVERNING EQUATIONS

All the governing differential equations of mass conservation, transport of momentum, energy, and mass fraction of species can be cast into a general partial differential equation as [25,26]

$$\frac{\partial \rho \varphi}{\partial t} + \frac{\partial}{\partial x_j} \rho \varphi u_j = \frac{\partial}{\partial x_j}\left(\Gamma \frac{\partial \varphi}{\partial x_j}\right) + S \qquad (2.1)$$

where φ is an arbitrary dependent variable (e.g., the velocity components and temperature), Γ is the generalized diffusion coefficient, and S is the source term for φ. The general differential equation consists of four terms. From the left to the right in Equation 2.1, they are referred to as the unsteady term, the convection term, the diffusion term, and the source term.

2.3 NUMERICAL SOLUTION METHODS FOR THE GOVERNING DIFFERENTIAL EQUATIONS

There are some methods established for numerical solution of the governing equations of fluid flow and heat transfer problems. They are the finite difference method (FDM) [27], the finite volume method (FVM) [25,26], the finite element method (FEM) [28,29], the control volume finite element method (CVFEM) [30], and the boundary element method (BEM) [31]. In this paper, only some details of the FVM will be presented.

2.3.1 THE FINITE VOLUME METHOD

In the FVM, the domain is subdivided into a number of so-called control volumes (CVs). The integral forms of the conservation equations are applied to each CV. At the center of the CV, a node point is placed. The variables are located at this node. The values of the variables at the faces of the CVs are determined by interpolation. The evaluation of the surface and volume integrals is carried out by quadrature formulas. Algebraic equations are obtained for each CV. Values of the variables for neighboring CVs appear in these equations.

The FVM is very suitable for complex geometries, and the method is conservative as long as surface integrals are the same for CVs sharing a boundary.

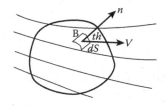

FIGURE 2.1 A control volume.

The FVM is a popular method especially for convective flow and heat transfer. It is also applied in several commercial CFD codes. Further details can be found in Patankar [25] and in Andersson et al. [26]. A brief illustration is presented in Equation 2.2, and an arbitrary CV is shown in Figure 2.1.

A formal integration of the general equation across the CV reads

$$\iiint_V \frac{\partial \rho U_j \phi}{\partial x_j} dV = \iiint_V \frac{\partial}{\partial x_j}\left(\Gamma_\phi \frac{\partial \phi}{\partial x_j}\right) dV + \iiint_V S_\phi\, dV \tag{2.2}$$

Then, by applying the Gaussian theorem or the divergence theorem, one has

$$\iint_S \rho \phi \bar{U} \cdot d\bar{S} = \iint_S \Gamma_\phi \nabla\phi \cdot d\bar{S} + \iiint_V S_\phi\, dV \tag{2.3}$$

By summing over all the faces of the CV, thus equation is transferred to

$$\sum_{f=1}^{nf} \phi_f C_f = \sum_{f=1}^{nf} D_f + S_\phi \Delta V \tag{2.4}$$

where the convection flux C_f, the diffusion flux D_f, and the scalar value of the arbitrary variable ϕ at a face, Φ_f, have to be determined.

2.3.1.1 Convection-Diffusion Schemes

To achieve physically realistic results and stable iterative solutions, a convection-diffusion scheme needs to possess the properties of conservativeness, boundedness, and transportiveness. The upstream, hybrid, and power-law discretization schemes all possess these properties and are generally found to be stable; but they suffer from numerical or false diffusion in multidimensional flows if the velocity vector is not parallel with one of the coordinate directions. The central difference scheme lacks transportiveness and is known to give unrealistic solutions at large Peclet numbers. Higher order schemes such as quadratic upstream interpolation convective kinetics (QUICK), van Leer, and so forth, may minimize the numerical diffusion but are less numerically stable. Also, implementation of boundary conditions can be somewhat problematic with higher order schemes; the computational demand can be extensive because additional grid points are needed and the expressions for the coefficients in Equation 2.4 become more complex.

2.3.1.2 Source Term

In general, the source term, S, may depend on the variable ϕ. In the discretized equation, it is desirable to account for such a dependence. Usually, the source term is expressed as a linear function of ϕ.

At the grid point P, S is then written as

$$S = S_C + S_P \varphi \qquad (2.5)$$

In order to prevent divergence, S_P must be negative, and the linearization procedure in Equation 2.5 is commonly used.

2.3.1.3 Solution of the Discretized Equations

The discretized equations resemble Equation 2.4 with the φ values at the grid points listed as unknowns. For boundaries not having fixed φ values, the boundary values can be eliminated by using given or fixed conditions of the fluxes at such boundaries. Gauss elimination is a so-called direct method for solving algebraic equations. For one-dimensional cases, the coefficients form a tridiagonal matrix, and an efficient algorithm (called the Thomas algorithm or the tridiagonal matrix algorithm [TDMA]) is achieved. For two-dimensional and three-dimensional cases, direct methods require large amounts of computer memory and computer time. Iterative methods are therefore used to solve the algebraic equations. One popular method is a line-by-line technique combined with a block correction procedure. The equations along the chosen line are solved by the TDMA. Iterative methods are also needed because the equations are nonlinear and sometimes interlinked.

In many situations (e.g., turbulent forced convection), the change in the value of φ from one iteration to another is so high that convergence in the iterative process is not achieved. To circumvent this and to reduce the magnitude of the changes, under-relaxation factors (between 0 and 1) are introduced.

2.3.1.4 The Pressure in the Momentum Equations

In the momentum equations, a pressure gradient term appears in each coordinate direction (i.e., a source term, S). If these gradients were known, the discretized equations for the flow velocities would follow the same procedure as for any scalar. However, in general, the pressure gradients are not known but have to be found as part of the solution. Thus, the pressure and velocity fields are coupled, and the continuity equation (mass conservation equation) has to be used to develop a strategy.

There are also other related difficulties in solving the momentum and continuity equations. It has been shown that if the velocity components and the pressure are calculated at the same grid points in a straightforward way, some physically unrealistic fields, such as checkerboard solutions, may arise in the numerical solution. A remedy to this problem is to use staggered grids. The velocity components are then given staggered or displaced locations. These locations are such that they lie on the CV faces that are perpendicular to them. All other variables are calculated at the ordinary grid points. Another remedy is to use a nonstaggered or collocated grid where all variables are stored at the ordinary grid points. A special interpolation scheme is then applied to calculate the velocities at the CV faces. Most often, the so-called Rhie–Chow interpolation method is applied [32].

2.3.1.5 Procedures for Solution of the Momentum Equations

As was mentioned in the preceding section, the velocity and pressure fields are coupled. Thus, a strategy has to be developed in the solution procedure of the momentum equations. The oldest algorithm is the semi-implicit method for pressure-linked equations (SIMPLE) algorithm. A pressure field is estimated and then the momentum equations are solved for this pressure field, resulting in a velocity field. Then a pressure correction and velocity corrections are introduced. From the continuity equation, an algebraic equation for the pressure correction can be obtained. The velocity corrections are related to the pressure corrections, and the coefficients linking the velocity corrections to the pressure correction depend on the chosen algorithm (see as follows). Then the momentum equations are solved again but with the corrected pressure replacing the guessed pressure. New velocities are obtained, and new pressure and velocity corrections are calculated. The whole process is repeated until convergence is obtained.

There are other similar algorithms available today, such as SIMPLE–consistent (SIMPLEC) and SIMPLE–extended (SIMPLEX). They differ from SIMPLE mainly in the expression for coefficients linking velocity corrections to the pressure correction; see Jang et al. [33].

Another, the pressure implicit split operator (PISO) algorithm [34], has become popular more recently. Originally, it served as a pressure–velocity coupling strategy for unsteady compressible flow. Compared to SIMPLE, it involves one predictor step and two corrector steps. Still another algorithm is SIMPLE–revised (SIMPLER). Here, the continuity equation is used to derive a discretized equation for the pressure. The pressure correction is then only used to update the velocities through the velocity corrections.

2.3.1.6 Convergence

The solution procedure is in general iterative; then some criterion must be used to decide when a converged solution has been reached. One method is to calculate residuals, R, as

$$R = \sum_{NB} a_{NB}\ \varphi_{NB} + b - a_{P}\varphi_{P} \tag{2.6}$$

for all variables. NB means neighboring grid points (e.g., E, W, N, S [east, west, north, and south]). If the solution is converged, $R = 0$ everywhere. Practically, it is often stated that the largest value of the residuals [R] should be less than a certain number. If this is achieved, the solution is said to be converged.

2.3.1.7 Number of Grid Points and Control Volumes

The widths of the CVs do not need to be constant nor do the successive grid points have to be equally spaced. Often, it is desirable to have a uniform grid spacing. Also, it is required that a fine grid be used where steep gradients appear, while a coarser grid spacing may suffice where slow variations occur. The various turbulence models require certain conditions on the grid structure close to solid walls.

The so-called high and low Reynolds number versions of these models demand different conditions.

In general, it is recommended that the solution procedure is carried out on several grids with differing fineness and varying degrees of nonuniformity. Then it might be possible to estimate the accuracy of the numerical solution procedure.

2.3.1.8 Complex Geometries

CFD methods based on Cartesian, cylindrical, or spherical coordinate systems have limitations in complex or irregular geometries. Using Cartesian, cylindrical, and/or spherical coordinates means that the boundary surfaces are treated in a stepwise manner. To overcome this problem, methods based on body-fitted or curvilinear orthogonal/nonorthogonal grid systems are needed. Such grid systems may be unstructured, structured or block-structured, or composite. Because the grid lines follow the boundaries, boundary conditions can more easily be implemented.

There are also some disadvantages with nonorthogonal grids. The transformed equations contain more terms, and the grid nonorthogonality may cause unphysical solutions. Vectors and tensors may be defined as Cartesian, covariant, contravariant, and physical or nonphysical coordinate-oriented. The arrangement of the variables on the grid affects the efficiency and accuracy of the solution algorithm.

Grid generation is an important issue. Today, most commercial CFD packages have their own grid generators. But several grid generation packages, compatible with some CFD codes, also are available. The interaction with various computer-aided design (CAD) packages is also an important issue today. Further information on treating complex geometries can be found in McBride et al. [35] and in Farhanieh et al. [36].

2.4 CFD APPROACH

The FVM described earlier is a popular method, particularly for convective flow and heat transfer. It is also applied in several commercial CFD codes. In heat transfer equipment such as heat exchangers both laminar and turbulent flows are of interest. Although laminar convective flow and heat transfer can be simulated, turbulent flow and heat transfer normally also require modeling approaches. In turbulence modeling, the goal is to account for all of the relevant physics by using as simple a mathematical model as possible. This section gives a brief introduction to the modeling of turbulent flows.

The instantaneous mass conservation, momentum, and energy equations form a closed set of five unknowns: u, v, w, p, and T. However, the computing requirements—in terms of resolution in space and time for direct solution of the time-dependent equations of fully turbulent flows at high Reynolds numbers (so-called direct numerical simulation [DNS] calculations)—are enormous, and major developments in computer hardware are needed. Thus, DNS is more often viewed as a research tool for relatively simple flows at moderate Reynolds numbers, with supercomputer calculations required. Meanwhile, practicing thermal engineers need computational procedures that supply information about the turbulent processes but avoid the need for predicting the effects of every eddy in the flow. This calls for information about the time-averaged properties of the flow and temperature fields (e.g., mean velocities, mean stresses, and mean temperature). Usually, a time-averaging operation, known as Reynolds decomposition, is carried out. Every variable is then written as a sum of a time-averaged value and a superimposed fluctuating value. In the governing equations, additional unknowns appear—six for the momentum equations and three for the temperature field equation. The additional terms in the differential equations are called turbulent stresses and turbulent heat fluxes, respectively. The task of turbulence modeling is to provide procedures for predicting the additional unknowns (i.e., the turbulent stresses and turbulent heat fluxes) with sufficient generality and accuracy. Methods based on the Reynolds-averaged equations are commonly referred as Reynolds-averaged Navier–Stokes (RANS) equation methods. Large eddy simulation (LES) lies between the DNS and RANS approaches in terms of computational demand. Like DNS, three-dimensional simulations are carried out over many time steps, but only the larger eddies are resolved. An LES grid can be coarser in space and the time steps can be larger than for DNS because the small-scale fluid motions are treated by a sub-grid-scale (SGS) model.

2.4.1 TURBULENCE MODELS

The most common turbulence models for industrial and heat exchanger applications are classified as

- Zero-equation models
- One-equation models
- Two-equation models
- Reynolds stress models (RSMs)
- Algebraic stress models (ASMs)
- LES models

The three first models in this list account for the turbulent stresses and heat fluxes by introducing a turbulent viscosity (eddy viscosity) and a turbulent diffusivity (eddy diffusivity). Linear and nonlinear models exist [37–39]. The eddy viscosity usually is obtained from certain parameters representing the fluctuating motion. A popular one-equation model is the Spalart–Allmaras model [40] where a transport equation is solved for the eddy viscosity. It is mostly used for aerospace and turbomachinery applications and less commonly for heat exchangers. In two-equation models, these parameters are determined by solving two additional differential equations. However, one should remember that these equations are not exact but approximate and involve several adjustable constants. Models using the eddy viscosity and eddy diffusivity approach are isotropic in nature and cannot evaluate nonisotropic effects. Various modifications and alternate modeling concepts have been proposed. Examples of models of this category are the k-ε, and k-ω models in high or low Reynolds number versions as well as in linear and nonlinear versions. A recently popular version is the V2F model introduced by Durbin [41]. It extends the use of the k-ε model by incorporating near-wall turbulence anisotropy and nonlocal pressure–strain effects while retaining a linear eddy viscosity assumption. Two additional transport equations are solved—one for the velocity fluctuation normal to walls and another one for a global relaxation factor. Recently, the shear stress transport k-ω model (SST k-ω) by Menter [42] has become popular because it uses a blending function of gradual transition from the standard k-ω model near solid surfaces to a high Reynolds number version of the k-ε model far away from a solid surface. Accordingly, it gives an accurate prediction of the onset and the size of separation under adverse pressure gradients.

In RSMs, differential equations for the turbulent stresses (Reynolds stresses) are solved, and directional effects are naturally accounted for. Six modeled equations (i.e., not exact equations) for the turbulent stress transport are solved together with a model equation for the turbulent scalar dissipation rate, ε. RSMs are quite complex and require large computing effort; for this reason, they are not widely used for industrial flow and heat transfer applications, such as for heat exchangers.

ASMs and explicit algebraic stress models (EASMs) present an economical way to account for the anisotropy of the turbulent stresses without solving the Reynolds stress transport equations. The idea is that the convective and diffusive terms are modeled or even neglected and then the Reynolds stress equations reduce to a set of algebraic equations.

For calculation of the turbulent heat fluxes, a simple eddy diffusivity concept (SED) most commonly is applied. The turbulent diffusivity for heat transport is then obtained by dividing the turbulent viscosity by a turbulent Prandtl number. Such a model cannot account for nonisotropic effects in the thermal field, but this model still is frequently used in engineering applications. Some models are presented in the literature to account for nonisotropic heat transport—for example, the generalized gradient diffusion hypothesis (GGDH) and the wealth equals earnings × time (WET) method. These higher order models require that the Reynolds stresses are calculated accurately by taking nonisotropic effects into account. If not, the performance may not be improved. In addition, partial differential equations can be formulated for the three turbulent heat fluxes, but numerical solutions of these modeled equations are rarely found. Further details can be found in Launder [43], for example.

In the LES model, time-dependent flow equations are solved for the mean flow and the largest eddies while the effects of the smaller eddies are modeled. The LES model is expected to emerge as the future model for industrial applications but it still limited to relatively low Reynolds numbers and simple geometries. Handling wall-bounded flows focusing on near-wall phenomena such as heat and mass transfer and shear at high Reynolds numbers present a problem due to near-wall resolution requirements. Complex wall topologies also present problems for LES.

Nowadays, approaches to combined LES- and RANS-based methods have been suggested. Called hybrid models, detached eddy simulation (DES) [44], is one such example.

2.4.2 WALL EFFECTS

There are two standard procedures to account for wall effects in numerical calculations of turbulent flow and heat transfer. One is to use low Reynolds number modeling procedures, and the other is to apply the wall function method. The wall function approach includes empirical formulas and functions linking the dependent variables at the near-wall cells to the corresponding parameters on the wall. The functions are composed of laws for the wall for the mean velocity and temperature and formulas for the near-wall turbulent quantities. The accuracy of the wall function approach increases with increasing Reynolds numbers. In general, the wall function approach is efficient and requires less CPU time and memory size, but it becomes inaccurate at low Reynolds numbers. When low Reynolds number effects are important in the flow domain, the wall function approach ceases to be valid. So-called low Reynolds number versions of the turbulence models are introduced, and the molecular viscosity appears in the diffusion terms. In addition, damping functions are introduced. Also, so-called two-layer models have been suggested to enhance the wall treatment. The transport equation for the turbulent kinetic energy is solved, while an algebraic equation is used for the turbulent dissipation rate, for example.

2.4.3   CFD CODES

Today, several industries, as well as engineering and consulting companies world-wide, are using commercially available general purpose CFD codes for simulation of flow and heat transfer topics in heat exchangers, investigations on enhanced heat transfer, electronics cooling, gas turbine heat transfer, and for other application areas (e.g., fuel cells). Among these codes are: ANSYS FLUENT, ANSYS CFX, STAR-CD, STAR CCM+, COMSOL, and PHOENICS. Also, many universities and research institutes around the globe apply commercial codes besides using codes developed in-house. Today, open-source codes such as OPEN-FOAM are available. However, to successfully apply such codes and to interpret the computed results, it is necessary to understand the fundamental concepts of computational methods. Other important issues are the description of complex geometries and the generation of suitable grids. The commercial codes typically have their own grid generation tool (e.g., ANSYS ICEM), but also stand-alone software such as Pointwise are popular. The codes are, generally, also compatible with CAD tools.

2.4.4   HOW TO ADOPT CFD FOR HEAT EXCHANGERS

CFD can be applied to heat exchangers in several ways. In one example, the entire heat exchanger or the heat transferring surface is modeled. This can be done by using large-scale or relatively coarse computational meshes or by applying a local averaging or porous medium approach. For the latter case, volume porosities, surface permeabilities, as well as flow and thermal resistances have to be introduced. The porous medium approach was first introduced by Patankar and Spalding [19] for shell-and-tube heat exchangers and has been followed up by many others.

Another method is to identify modules or group of modules that repeat themselves in a periodic or cyclic manner in the main flow direction. This will enable accurate calculations to be made for the modules, but the entire heat exchanger (including manifolds and distribution areas) is not included. The idea of stream-wise periodic flow and heat transfer was introduced by Patankar et al. [45].

2.5 POROUS MEDIA HEAT EXCHANGERS

Graphite foam is a favorable porous material in thermal engineering applications because of its high thermal conductivity (the solid phase thermal conductivity is 1500–2000 W/[m·K], and the effective thermal conductivity is varied with a density from 40 to 150 W/[m·K]) and the large specific surface area (5000–50,000 m²/m³) [46]. Graphite foam is a favored material in thermal engineering applications such as electronic cooling systems, vehicle cooling systems, energy storage systems, high-temperature heat exchangers, and so on. However, there is an associated high flow resistance in graphite foam that results from the porous structure property. In order to reduce the flow resistance and enhance the heat transfer, dimpled fins could be applied in graphite foam heat exchangers. Reviews of graphite foam used in thermal engineering applications are available [47].

Even though there is a substantial heat transfer enhancement in graphite foam, it is associated with other problems. An important issue in graphite foam use is the high pressure drop due to the large hydrodynamic loss associated with the cell windows connecting the pores [48]. In a study concerning reduction of the pressure drop, Gallego and Klett [49] presented six different configurations of graphite foam heat exchangers. That study showed that solid foam had the highest pressure drop and a finned configuration had the lowest. In another study, Leong et al. [50] found that baffle foam presented the lowest pressure drop among four configurations of graphite foams at the same heat transfer rate. Lin et al. [51] revealed that corrugated fins could reduce the pressure drop while maintaining a high heat transfer coefficient compared to solid foam. All together, these studies illustrate that the configuration has an important effect on the pressure drop through the graphite foam.

2.5.1 CASE STUDY: FOAM HEAT EXCHANGER

From the aforementioned literature review, it is evident that dimple technology may enhance heat transfer in a channel or in a heat exchanger. However, the available research is limited concerning the dimple technology used in porous graphite foam, which is a very good potential material for CHEs.

In this chapter, the flow characteristics and thermal performance of graphite foam dimpled fin heat exchangers will be considered with the application aimed at vehicle radiators. Three-dimensional numerical simulations of fluid flow and heat transfer in graphite foam dimpled fin channels will be presented. In order to take advantage of the graphite foam's high thermal conductivity and low density and also to reduce the foam's flow resistance, fins with one-sided dimples or two-sided dimples are designed into the graphite foam heat exchanger. The main objectives are to demonstrate the computational results used to investigate the heat transfer enhancement of a graphite foam dimple fin and the flow characteristics caused by the porous graphite foam dimple fin. A local thermal nonequilibrium model is applied to analyze the thermal performance of the graphite foam dimple fin (porous zone), and the Forchheimer extended Darcy's law has been used to consider the air pressure drop through the porous graphite foam. In addition, the SST k-ω turbulence model has been used to capture the turbulent flow characteristics outside the graphite foam region.

The details of the fluid flow and heat transfer over the dimple fin will be presented later in this chapter.

Figure 2.2 shows the schematics of the considered dimpled fin heat exchanger. A core of graphite foam fin is chosen with only two rows of dimples (x-direction) and half of the fin width; the overall size of the core is 16 mm × 32 mm × 15 mm (x × y × z). Figure 2.3 shows more details of the dimple fin geometry. The imposed boundary conditions are illustrated in Figure 2.4.

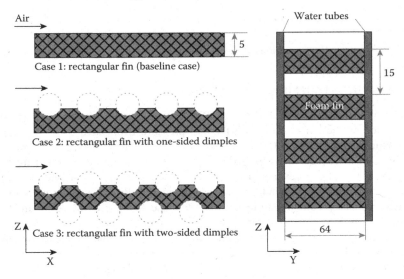

FIGURE 2.2 Schematic figures of the dimpled fin heat exchangers. Dimensions are in mm.

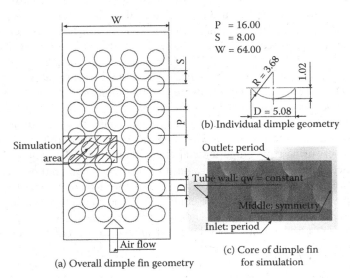

FIGURE 2.3 Schematic figures of the dimpled fin geometry. Dimensions are in mm.

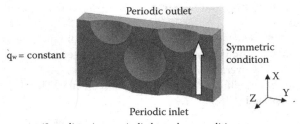

Periodic outlet

$\dot{q}_w$ = constant

Symmetric
condition

X

Z   Y

Periodic inlet

*In z-direction: periodic boundary condition

FIGURE 2.4   Boundary conditions of the dimple fin cases.

2.6   ARTIFICIAL NEURAL NETWORKS

In order to evaluate the performance of heat exchangers, efficient and accurate methods for prediction of heat transfer and pressure drop have to be developed. Computational intelligence (CI) techniques, such as ANNs, GAs, and fuzzy logic (FL), have been successfully applied in scientific research and engineering practices. ANNs have been developed for several decades and are now widely used in various application areas—such as in performance prediction, pattern recognition, system identification, and dynamic control—because ANNs provide better and more reasonable solutions. ANNs offer new ways to simulate nonlinear, uncertain, or unknown complex systems without requiring any explicit knowledge about input/output relationships. An ANN has more attractive advantages. It can approximate any continuous or nonlinear functions by using a certain network configuration. It can be used to learn complex nonlinear relationships from a set of associated input/output vectors. It can be implemented to dynamically simulate and control unknown or uncertain processes. ANNs have been used in thermal systems for heat transfer analysis, performance prediction, and dynamic control of heat exchangers [52–67]. For example, Yang and Sen [52,53] reviewed works on dynamic modeling and control of heat exchangers using ANNs and GAs. Two interesting examples were presented to support the superiority of ANNs and GAs compared to correlations. Diaz et al. [54–57] did lots of work in steady and dynamic simulation and control of a single-row fin-and-tube heat exchanger (FTHE) using ANNs. Pacheco-Vega et al. [58,59] performed heat transfer analysis for a fin-tube heat exchanger based on limited experimental data with air and R22 as fluids and predicted heat transfer rates of air-water heat exchangers using soft computing and global regression. Islamoglu et al. [60,61] predicted heat transfer rates for a wire-on-tube heat exchanger and outlet temperature and mass flow rate for a nonadiabatic capillary tube suction line heat exchanger. Based on their experimental data, Xie et al. [62,63] conducted heat transfer analysis and performance prediction of shell-and-tube heat exchangers with helical baffles. Ertunc et al. [64–66] conducted neural network analysis and prediction for an evaporative condenser, a cooling tower, and a cooling coil. Zdaniuk et al. [67] combined their data and other databases to predict the performance of helically finned tubes by a single-output ANN. Besides heat exchanger applications, there have been other uses of ANNs for heat transfer analysis and fluid flow process predictions [68–74].

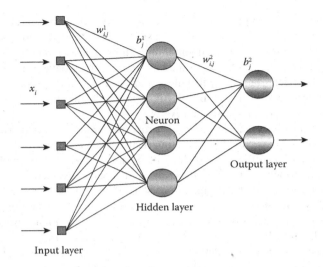

FIGURE 2.5 Illustration of artificial neural network (ANN): input layers, one or more hidden layers, weights, bias, transfer functions, and output layers.

From the aforementioned successful applications, it is has been shown that ANNs are efficient techniques for estimation or prediction of heat transfer and fluid flow processes and are well suited to thermal analysis in engineering systems, especially in heat exchangers. An illustration of an ANN is shown in Figure 2.5.

Implementation of an ANN into heat exchanger design codes can be performed to facilitate the prediction of thermal and hydraulic coefficients. The ANN model can accurately predict heat transfer and friction factor. This can then be applied to solve an inverse heat transfer problem via optimization.

In the aforementioned literature, most thermal analysis focused on FTHEs with small numbers of tube rows (usually fewer than four) or small tube diameters (mostly in the range of 8–13 mm), which are commonly used in heating, ventilation, air conditioning, and refrigeration (HVAC&R) engineering or in shell-and-tube heat exchangers [54–59]. However, ANNs have not yet been applied to correlate and predict heat transfer and flow friction of FTHEs with large numbers of tube rows and large tube diameters. In some large industry applications, such as intercoolers of multistage compressors used in high pressure and temperature plants (such as gas turbine plants), the number of tube rows might be greater than four or even much larger. The outside diameter of the tubes might be larger than 13 mm or even much larger). The intercooler is used to cool the air at the compressor intake, resulting in a decrease in the work of the compressor and an increase in the network from the plants. Most of the aforementioned ANN configurations only have a single output [60–63,67–74], implying that more sets of networks should be developed for different outputs. For this reason, the objective of this chapter is to present ANN development and applications for heat transfer analysis of FTHEs having large tube diameters and large numbers of tube rows, with experimental data being used in the back propagation algorithm for training the network.

Different network configurations have been considered in the search for an optimal network configuration for prediction. Next, the ANN was trained again with the combined database from the experimental data and numerical data by CFD for turbulent cases. Finally, the ANN was trained by the combined database of experimental data and numerical data from turbulent and laminar computations. The results extend those presented by Xie et al. [75,76], where only the turbulent heat transfer and fluid flow were correlated using an ANN.

Improving heat exchanger design is an active research field for thermal designers. There are many experimental and numerical studies on heat exchanger design. However, generally it is difficult to collect all the data for design because of the extensive costs of experiments (including those for setups and manufacturing of samples) and the computer resources for computations. Therefore, the use of different methods can be considered as alternative techniques for design. The ANN approach is useful and convenient for engineers or researchers to predict the performance of a given heat exchanger with limited experimental data. It does not need to provide an accurate and detailed mathematical formulation. Once the ANN has been trained, the weights and biases from the network, which correspond to a practical heat exchanger, can be transferred to engineers or researchers who are going to use the test data for prediction. Then engineers may simply feed these data into the trained network and quickly make accurate predictions of the thermal performance for the practical heat exchanger. However, some limitations should be considered when using ANNs because they do not provide any knowledge about the physical phenomena and do not correlate the information. This means that an ANN does not know the inherent physical principles and how and why the phenomena occur, change, or disappear.

A review on applications of ANN for thermal analysis of heat exchangers was recently presented by Mohanraj et al. [8].

2.7 GENETIC ALGORITHMS

In recent years, applications of GAs in thermal engineering have received much attention for solving real-world problems [52]. The application of GAs into heat exchanger optimizations has revealed that GAs have a strong ability to successfully optimize and predict thermal problems. Thus, applications of GAs in the field of thermal engineering present new challenges. At this point, GAs may be used in the geometrical optimization of heat exchangers in order to obtain optimal results under specified design objectives within allowable pressure drops. For example, FTHE performance has been predicted using a GA [77]. Shell-and-tube heat exchangers are optimized under four strategies by using simulated annealing (SA) and GA, and the corresponding performances are compared [78]. A new design method was proposed to optimize a shell-and-tube heat exchanger from an economic point of view by a GA [79]. Optimization of the geometry of cross-wavy (CW) and cross-corrugated (CC) primary surface recuperators was studied via GAs [80,81]. Heat transfer correlations for CHEs were obtained using GAs and, in turn, these correlations were used to estimate performance [82–84].

In addition, plate-fin heat exchangers were optimized by means of GAs [84–86]. Mishra et al. [84] used a GA to optimize a multilayer plate-fin heat exchanger in which the specified heat duty and flow restrictions were alternately considered and the effects of additional constraints on optimum design were investigated. Ozkol et al. [85] determined the optimum geometry of a heat exchanger by a GA in which the ratio of the number of heat transfer units (NTUs) to the airside pressure drop was the objective function, and two examples demonstrated the GA method. Xie and Wang [86] optimized a plate-fin heat exchanger by using a GA. The fin geometries were scaled to be searched and optimized for minimum total weight or maximum effectiveness.

The aforementioned research draws attention to geometrical optimization of plate-fin heat exchangers by GAs. However, although the pressure drop constraints are clearly considered in several studies [84–86], the objective of minimum total annual cost is not included in some of these [85,86]. Generally, the optimal design of the heat exchanger is conducted under the constraint of an allowable pressure drop, although the contradictory case (without pressure drop constraints) is not often considered. In some applications, the pressure drop is not the most critical parameter. It might be that the gain/benefit of the optimized performance is worthwhile even if a pressure drop penalty is included. Therefore, in this chapter, an optimal design for CHEs with or without constraints on the allowable pressure drop is considered using GA auto-search, combination, and optimization techniques.

On the other hand, for heat exchanger optimization, the trade-off between heat transfer and pressure drop may be considered. In general, a higher flow velocity means a higher heat transfer coefficient and hence a smaller heat transfer area and correspondingly lower capital cost. However, higher velocity generally will lead to a higher pressure drop and hence a higher power consumption and correspondingly higher power cost. The heat exchanger area and the pressure drop are mainly associated with capital cost (investment cost) and power cost (operating cost), respectively. Thus, the important objective—minimum cost—should be considered ahead of the optimum design. In addition, in some practical applications, where the CHE size/volume is critical for compactness and the total cost should be low, moderate volume and cost must be considered for the specified requirements; usually equal consideration is not given to their importance.

In this chapter, a GA-based optimization technique is demonstrated for a plate-fin CHE. The objectives of the minimum total volume and/or total annual costs of the heat exchangers are considered for optimization. The fin geometries are taken from an available database, and three shape parameters are varied for the optimization objectives with or without pressure drop constraints. The considered method, GA, is not new; however, the application of GA into plate-fin heat exchanger optimizations for different objectives with/without different constraints is attractive, and the results are optimistically interesting, both of which are useful for further research or references.

Recently, a relevant study applying a GA for optimization of a finned-tube heat exchanger modeled with the volume averaging theory was presented by Geb et al. [9].

2.7.1 SIMPLE DESCRIPTIONS

The GA is maintained by a population of parent individuals that represent the latent solutions of a real-world problem. For example, the designer might encode the design parameters into corresponding binary strings; then all the binary strings

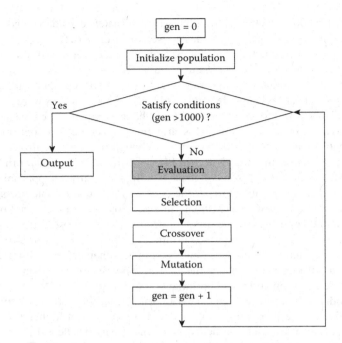

FIGURE 2.6　Flow chart of the genetic algorithm (GA)—an optimization method.

are connected into a single binary string, which is represented as an individual. Accordingly, a certain number of design parameter sets become a population of parent individuals. Figure 2.6 shows the flow chart of the GA. Each individual is assigned a fitness based on how well each fits a given environment, and then each individual is evaluated by survival of that fitness. Fit individuals go through the process of *survival selection*, *crossover*, and *mutation*, resulting in the creation of the next generation—called child individuals. Therefore, a new population is formed by selection of good candidates from the parent and child individuals. After some generations, the algorithm has reached convergence to a best individual, which probably represents the best solution of the given problem. More details and descriptions of GAs can be found in various textbooks [87–89].

2.7.2 Ranges of the Variables

A binary string is adopted for encoding the variables of a given model. The search ranges and binary string length of the design parameters must be specified. The selection of upper-bound and lower-bound shape parameters can be based on references or on practical constraints. Considering the ability of computer handling bit operation and engineering application, the computational precision must be set. The total number of possible combinations is usually very high, and this means that it will require enormous time and effort until the global optimization solution is reached. The detailed principles of the coding and decoding processes can be found in the aforementioned classical textbooks [87–89].

2.7.3 GENETIC OPERATORS AND PARAMETERS

In this section, tournament selection, uniform crossover, and one-point mutation are selected. And niching (sharing) and elitism are adopted [90–92].

Tournament selection: Random pairs of individuals are selected from the population under a given probability, and the better (those with greater fitness) of each pair is allowed to mate; thus one child is created, which becomes a mix of the two parent individuals.

Uniform crossover: It is possible to obtain any combination of two parent individuals (e.g., considering two individuals, 011100 and 101011, the children could be 111010 and 001001).

One-point mutation: There is a small probability that one or more of the children will be mutated (e.g., 111010 could be 101010).

Niching (sharing): The fitness of each individual is adjusted according to a similar degree, which is evaluated by a specified sharing function. The selection is conducted based on the new fitness. This technique ensures the variety of individuals (solutions) so that the global, rather than the local, optimization solution will be obtained.

Elitism: The best parent is reproduced (copied) into the new population. After the new population is generated, the GA checks whether the best parent has been replicated.

The selection of genetic parameters is a trial-and-error process, and with the variation of these parameters, results are not exactly identical. Optimization of heat exchangers and design objective definitions will be discussed in another section.

2.8 CFD: APPLICATION TO PLATE HEAT EXCHANGERS

As the heat transferring plates in plate heat exchangers (PHEs) are designed, knowledge about mass flow, pressure loss, and temperature distribution in the channels is important. Various types of PHEs exist, but generally they are made up of plates with an embossed surface area enhancement pattern. Figure 2.7 shows a plate-and-frame PHE. A common pattern is the chevron or herringbone design. As the plates are assembled, complex and narrow channels appear. Since the beginning of PHE production, the foundation for new designs has been experimental results. Researchers and developers of PHEs have formulated empirical relationships for the main flow and thermal properties, that is, pressure drop (friction factors) and heat transfer coefficients. Until recently, these empirically based design methods have been sufficient because of the high performance of the PHE. Today, however, there is a growing demand for process-adjusted heat exchangers that are not based solely on overall experimental data. This has been paired with demands for more advanced design methods due to higher energy costs (i.e., design and optimization of CHEs).

The different channels are commonly separated by a sealing gasket. Development of new technologies has resulted in brazing or welding plates together. Welded PHEs are classified as fully welded or semi-welded. The former do not use any

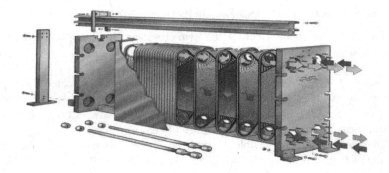

FIGURE 2.7   Plate heat exchanger. (Courtesy Alfa Laval AB.)

FIGURE 2.8   Plate with chevron pattern.

elastomer sealing between the plates, while in the latter, pairs of plates are welded together. For the semi-welded PHEs, every pair of plates makes up a cassette in which one fluid flows. The other fluid flows among neighboring cassettes that are sealed with gaskets.

Figure 2.8 shows a plate with a chevron pattern. Plates are usually stacked together in a symmetric or mixed arrangement as indicated in Figure 2.9.

Investigations of PHE flow fields have been carried out only to a limited extent—mainly because of the complex geometry and narrow passages having small hydraulic diameters. The conjectured flow patterns in the washboard type of corrugated plate and in the herringbone type of plate are shown in Figure 2.10a and b, respectively.

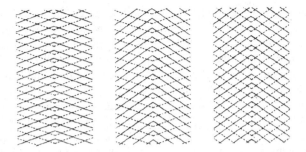

FIGURE 2.9 Sketch of plate arrangement.

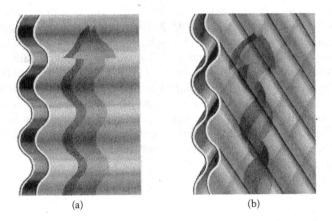

FIGURE 2.10 (a) Washboard pattern plates. (b) Herringbone pattern plates.

PHEs are used in many different processes with a broad range of temperatures and a variety of substances. Various types of PHEs exist, and one example is shown in Figure 2.7. PHEs are all assemblies of plates with an embossed surface area enhancement pattern. Today, the most common patterns in use are the chevron or herringbone designs. The embossed pattern results in a larger heat transfer surface than that of a flat plate, improves the plate stiffness, and assures the channel gap. The geometry of the corrugation is very important for the thermal-hydraulic performance of the heat exchanger. Each plate has four corner ports that, in pairs, establish access to the narrow flow passages on either side of the plate. Usually, there are distribution areas between the ports and the major embossed plate area.

CFD can be applied to PHEs in different ways. In one method, the entire heat exchanger or the heat transferring surface is modeled. This can be done by using large-scale or relatively coarse computational meshes or by applying a local averaging or porous medium approach. For the latter case, volume porosities, surface permeabilities, as well as flow and thermal resistances have to be considered. The porous medium approach was first introduced by Patankar and Spalding [19] for shell-and-tube heat exchangers and many other studies have followed.

Another way to apply CFD is to identify modules or group of modules that repeat themselves in a periodic or cyclic manner in the main flow direction. A unitary cell usually can be identified for PHEs. This will enable accurate calculations for the modules, but the entire heat exchanger—including manifolds and distribution areas—is not included. The idea of stream-wise periodic flow and heat transfer was introduced by Patankar et al. [45] but not directly for PHEs.

In this chapter, the simulated thermal and hydraulic characteristics of corrugated fluid channels of compact brazed PHEs by application of the commercial CFD software ANSYS CFX 14.0 are presented. The influence of the geometry parameters of the corrugated pattern (such as chevron angle and corrugation pitch) on the brazed plate heat exchanger (BPHE) performance have been studied. The influence of various types of wall heat transfer boundary conditions on the simulation results have also been investigated. An entire fluid channel was simulated using various turbulence models in the Reynolds number range of 300–3000. The reported results were calculated by using the Launder–Reece–Rodi isotropic production RSM. The CFD predictions were validated using available experimental data. The computational domain representing the entire fluid channel is shown in Figure 2.11a. An unstructured tetrahedral mesh with prism layers (created by ANSYS ICEM software) is illustrated in Figure 2.11b.

Figure 2.12 presents the heat transfer predictions (Nu/Pr$^\gamma$ versus Reynolds number) for various thermal boundary conditions. It is revealed that using an external heat transfer coefficient with the ambient temperature as the thermal boundary condition results in superior behavior. Figure 2.13 demonstrates flow patterns for two different

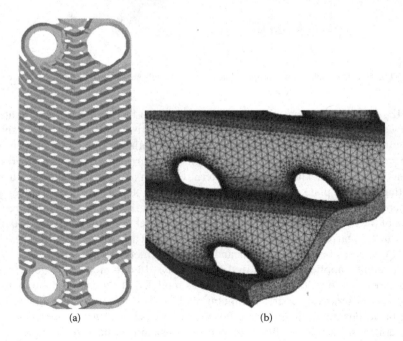

(a)   (b)

FIGURE 2.11   (a) Corrugated pale fluid channel with indicated braze junctions. (b) Unstructured tetrahedral mesh with prism layers.

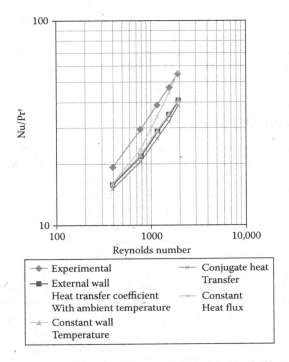

FIGURE 2.12 Comparison of thermal performance using various thermal boundary conditions.

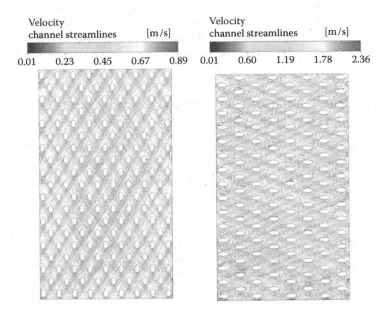

FIGURE 2.13 Flow patterns for chevron angles 32° and 67° at inlet Reynolds number of 2000.

chevron angles at a Reynolds number of 2000. The importance of the chevron angle for the flow field is revealed.

Further details of the computations for brazed PHEs can be found in the papers by Gullapalli and Sunden [93,94], while Etemad and Sunden [95] present recent simulations for the unitary cell of a larger PHE.

2.9   DUCTS WITH BUMPS

A duct with bumps is considered as another example. This type of duct appears in some rotary regenerative heat exchangers in operational heat recovery systems. The basic concept of introducing bumps is to design corrugated ducts as indicated in Figure 2.14 for a triangular cross section. The intention is that this corrugation should affect the flow field and introduce low Reynolds number turbulence and a swirling motion as conjectured in Figure 2.14. At a certain distance downstream, the corrugation element, the turbulence, and the swirling motion will be attenuated, and gradually, the intensity of the fluctuations will be reduced. Therefore, at a position upstream where the complex flow pattern (strong secondary cross-sectional flow and separated flow) has been significantly weakened or has disappeared, a new corrugation element is introduced to reestablish a violent swirling motion. CFD calculations have been performed and a nonorthogonal structured grid was utilized. Periodic conditions were imposed in the main flow direction. About 40,000 CVs were used, with 30×60 CVs in the cross-sectional plane. The existence of a secondary flow was revealed, and a result is shown in Figure 2.15. It is obvious that a swirling motion is created by the bumps and the triangular cross section. The Reynolds number corresponding to the flow in Figure 2.15 is about 2000. A low Reynolds number k-ε model was used in the simulations. The secondary motion also exists for laminar cases because they are partly geometry driven. The heat transfer is enhanced compared to a smooth duct, but the pressure drop increase is high. Further details of this investigation can be found in Sunden [96].

2.10   CFD: APPLICATION TO GRAPHITE FOAM DIMPLED FIN HEAT EXCHANGERS

In this section, some results for the case study described in Section 2.5.1 are presented. A schematic picture of the physical model of the plate-fin heat exchanger is shown in Figure 2.2. Hot water flows inside the flat tubes, and the cold air flows

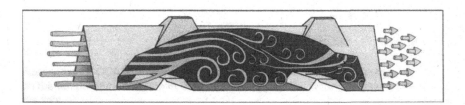

FIGURE 2.14   Conjectured flow pattern in a duct with bumps.

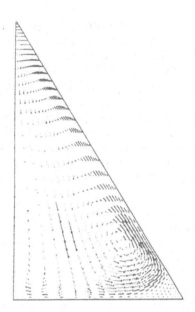

FIGURE 2.15 Secondary flow velocity vectors in a cross-sectional plane midway over a corrugation element.

through the porous graphite foam fins. The heat is transmitted through the tube wall to the graphite foam fins and finally is dissipated into the air. Three cases of graphite foam fins are analyzed: rectangular fin without dimple (Case 1); rectangular fin with one-sided dimples (Case 2); and rectangular fin with two-sided dimples (Case 3). Note that the fin thickness of the three cases is kept the same. Accordingly, in this instance, the thickness of the graphite foam fin selected is 5 mm. The pitch of the fins in the height direction (z-direction) is 15 mm, and the width of the fins is 64 mm. The assumptions are as follows:

1. The air is assumed to be incompressible, with constant properties and a steady state.
2. The connection between the tube wall and the graphite foam fin is assumed to be perfect, without any air gap inside. Thus, the thermal resistance at the interface between the tube wall and the graphite foam is neglected.
3. The porosity through the graphite foam dimpled fin is constant.
4. The thermal conductivity of the graphite foam is assumed to be isotropic.

The commercial code ANSYS FLUENT 14.0 was used for the numerical simulations. The SIMPLE algorithm was used to couple pressure and velocity. A second-order upwind scheme was used for the space discretization of the momentum and energy equations in the simulations. In order to achieve convergent results, the residual of the continuity, components of velocity, k, ω, intermittency, and Reynolds number based on momentum thickness was set to be below 10^{-4}, while for the energy it was below 10^{-7}.

Due to the complex geometries, an unstructured grid is developed using ICEM software, as shown in Figure 2.16. To ensure the accuracy and validity of the numerical results, grid independence tests were conducted. Four sets of mesh size (Mesh 1: 225,000; Mesh 2: 696,000; Mesh 3: 1,177,000; and Mesh 4: 2,000,000) were tested with Case 2 at a Reynolds number of 8536. The y^+ – values range from 0.2 to 0.7, which indicates that the grid was sufficiently fine.

Details of the flow streamlines inside the dimple are shown in Figure 2.17. The recirculating flow, which occurs upstream of the dimple, is a three-dimensional phenomenon. Due to the strong reattachment of the shear layer, some part of the recirculating flow traverses the side of the dimple, loses its momentum, and ejects along the side rim of the dimples.

The overall performance criterion $\overline{Nu}/\overline{Nu_0}/(f/f_0)^{1/3}$, is commonly used as various configurations are compared. This performance parameter provides a heat transfer augmentation quantity under conditions of a certain input pumping power and a certain heat transfer duty. This parameter thus considers both the heat transfer augmentation and the friction loss increase. Figure 2.18 presents values of this parameter for the three considered cases. The graphite foam fin with two-sided dimples provides the highest value of $\overline{Nu}/\overline{Nu_0}/(f/f_0)^{1/3}$, which is between 1.5 and 2.7. This implies that the graphite foam fin with two-sided dimples can enhance the heat transfer maximum 2.7 times under the same input pumping power and the same heat transfer duty compared with the other two cases. Therefore, the graphite foam fin with two-sided dimples exhibits good overall performance.

In terms of energy savings, the graphite foam fin with one or two-sided dimples provides higher performance than the conventional aluminum offset fin, wavy fin, and louvered fin. Furthermore, the graphite foam fin with one-sided dimples presents higher effectiveness in energy savings than one with two-sided dimples.

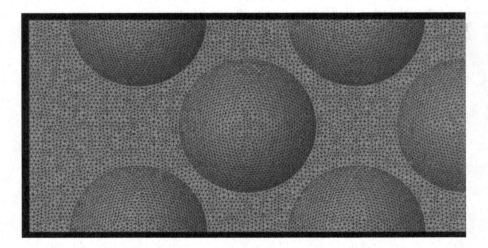

FIGURE 2.16   Unstructured grid.

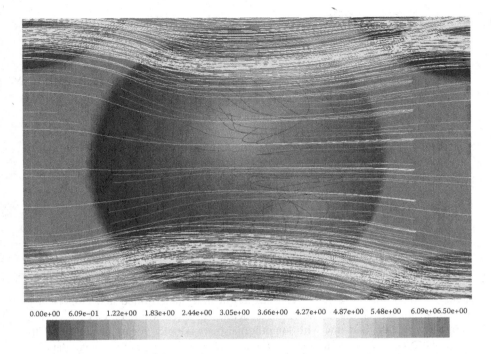

0.00e+00 6.09e−01 1.22e+00 1.83e+00 2.44e+00 3.05e+00 3.66e+00 4.27e+00 4.87e+00 5.48e+00 6.09e+06.50e+00

FIGURE 2.17 Velocity field (colored by velocity), Re = 5687. Areas of sudden expansion, negative pressure, and recirculating flow can be identified.

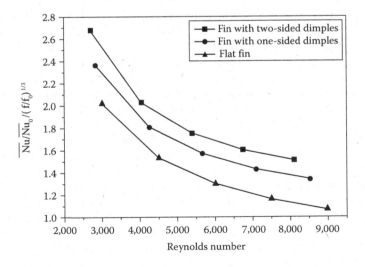

FIGURE 2.18 Overall thermal performances for three sets of foam fins.

The graphite foam fin with dimples has the potential to enhance the heat transfer.

1. The dimples lead to strong flow mixing along the graphite foam fin and enhance the heat transfer.
2. The graphite foam fin with two-sided dimples enhances heat transfer up to 4.6 times. However, due to the porous structure of the graphite foam, the normalized friction factor, f/f_0, is between 5.2 and 4.0.
3. Combining the enhanced heat transfer and pressure drop increase, the graphite foam fin with two-sided dimples presents the highest value in the overall performance criterion, that is, $Nu/Nu_0/(f/f_0)^{1/3}$.

Further details of recent research on graphite foam fin heat exchangers can be found in Lin et al. [97–99].

2.11   RESULTS OBTAINED BY ANN AND GA

2.11.1   ANN

Here, an ANN was used to correlate experimentally determined and numerically computed Nusselt numbers and friction factors for three kinds of FTHEs having plain fins, slit fins, and fins with longitudinal delta-winglet vortex generators with large tube-diameter and a large number of tube rows. First, the experimental data for training the network were picked up from a database of nine samples with a tube with an outside diameter of 18 mm; six, nine, and twelve tube rows; and Reynolds numbers that varied between 4000 and 10,000. The ANN configuration under consideration had twelve inputs of geometrical parameters and two outputs of heat transfer Nusselt number and fluid flow friction factor. The commonly implemented feed-forward back propagation algorithm was used to train the neural network and modify weights. Different networks with various numbers of hidden neurons and layers were assessed to find the best architecture for predicting heat transfer and flow friction. The deviation between the predictions and experimental data was less than 4%. Compared with correlations for prediction, the performance of the ANN-based prediction exhibited superiority. Next, the ANN training database was expanded to include experimental data and numerical data of other similar geometries by CFD for turbulent and laminar cares with Reynolds numbers of 1000–10,000. This in turn indicated that the predicted data was in good agreement with the combined database. The satisfactory results suggest that the developed ANN model was generalized to predict the turbulent or laminar heat transfer and fluid flow through three such kinds of heat exchangers with large tube diameters and large numbers of tube rows.

In order to predict heat transfer and friction with high precision, attempts should be made to develop some ANN configurations and finally to find a relative optimal configuration for prediction. As mentioned earlier, the drawback of the BP algorithm is that it may get stuck in a local minimum; therefore, the learning rate was changed during the training process of the network. Various neural network configurations were tested based on experimentally measured databases of Nusselt numbers and friction factors for three kinds of heat exchangers. It was concluded that the 12-9-5-2

(12 input layers, 2 hidden layers with 9 and 5 nodes, respectively, and 2 output layers) feed-forward neural network was the most accurate architecture for prediction of turbulent heat transfer and flow resistance for the FTHEs having plain fins, slit fins, and fins with longitudinal vortex generators with large tube diameters and large numbers of tube rows. It was found to provide errors within 2% at the end of the training process. Also, the ANN yielded superior prediction of heat transfer and flow friction compared to power-law or multiple correlations.

By extension, to train the ANN for the combined database from experimental data and numerical data by the CFD technique, the developed ANN architecture became more generalized and universal. Therefore, the ANN 12-9-5-2 architecture showed a strong ability to predict heat exchanger performance for turbulent and laminar heat transfer and fluid flow with an accepted deviation close to the measurement uncertainty/error.

In conclusion, it must be remembered that the limitation of ANN is that it cannot describe the unknown physical phenomena directly. Further details from the present ANN studies can be found in the literature [100,101].

2.11.2 GA

In this section, a plate-fin type CHE was considered for optimization. The optimization method used a GA to search, combine, and optimize structure sizes of the CHE. The minimum total volume and/or total annual costs of PFCHE were taken as objective functions in the GA. The geometry of the fins was fixed while three shape parameters were varied for the optimization objectives with or without pressure drop constraints. The implementation of the design method consisted of a GA routine and a rating performance routine. In the GA routine, binary coding for tournament selection, uniform crossover, and one-point mutation was adopted, and niching (sharing) and elitism were implemented. In the rating routine, performance of the CHE was evaluated according to the conditions of the structure sizes that the GA generated, and the corresponding volume and cost were calculated. The results showed that the optimized CHE provided lower volume and/or lower annual cost than those presented in the literature. The method is universal and may be used for optimization of new CHEs under different objectives.

In this case, four objective functions were considered in the design process. The first was the minimum total cost (C), the second was the minimum total volume (V), the third was moderate cost and volume (CV), and the last one was the minimum total pressure drop (DP). The traditional rating method, the $\varepsilon - NTU$ method, was used in the rating performance of the PFCHE.

Here, the GA optimization with pressure drop constraints was called GA1 while GA optimization without pressure drop constraints was called GA2. As an example, optimization was carried out using a PFCHE heat exchanger with the following duty: 0.6 m³/s air at 4°C was cooling 1.2 m³/s gas from 240°C to 51°C. The inlet pressure was 110 kPa. The plate thickness was 0.4 mm, and both fins and plates were made from aluminum. Plain triangular fins and offset strip fins were used on each side. The evolution process for minimum cost is shown in Figure 2.19. The impact of the pressure drop constraint is evident.

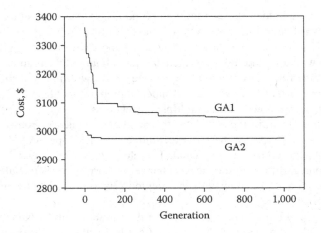

FIGURE 2.19 Evolution process for minimum cost.

It was concluded that the GA has a strong ability for auto-search and combined optimization in the optimization design of heat exchangers compared to the traditional designs in which a trial-and-error process may be involved. By application of the GA in the optimal design the heat exchanger, configurations/structures can be optimized according to different design objectives, such as minimum surface area and cost. The method can be transferred for use in optimization design of different types of heat exchangers, but it can also be used for the presented plate-fin heat exchangers but with different fins (e.g., perforated fins, slotted fins, and louvered fins).

Further details of GA applications for the design and optimization of heat exchangers can be found in the literature [101–104].

2.12 TOPICS NOT TREATED IN THE CFD DESCRIPTION

There are several additional topics that are of importance in CFD modeling and simulations of heat exchangers. Among those not being treated in this chapter are:

- Implementation of boundary conditions
- Adaptive grid methods
- Local grid refinements
- Solution of algebraic equations
- Convergence and accuracy
- Parallel computing
- Animation

2.13 CONCLUSIONS

Various computational approaches for analysis of transport phenomena were briefly summarized and reviews of recent works were provided. Application of computational methods (e.g., CFD) for complex fluid flow and heat transfer were described for important engineering issues related to PHEs, radiators, and heat recovery units. Each application prohibits important physical and computational aspects. The results have revealed that the CFD approach can demonstrate important physical effects as well as provide satisfactory results in decent agreement with corresponding experiments in terms of, for example, Nusselt numbers and pressure drop with respect to the friction factor.

Neural network configurations were introduced, and a case study was illustrated where the neural network was tested based on experimentally measured databases of Nusselt numbers and friction factors for three kinds of heat exchangers. It was concluded that the 12-9-5-2 feed-forward neural network was the most accurate architecture for prediction of turbulent heat transfer and flow resistance for the FTHEs having plain fins, slit fins, and fins with longitudinal vortex generators with large tube diameters and large numbers of tube rows. Also, the ANN presented yielded superior prediction of heat transfer and flow friction compared to power-law or multiple correlations. By extension, in order to train the ANN for the combined database derived from experimental data and numerical data by the CFD technique, the developed ANN architecture can be altered to be more general and universal. Therefore, the presented ANN 12-9-5-2 architecture showed a strong ability to predict the heat exchanger performance of turbulent and laminar heat transfer and fluid flow with an accepted deviation close to the measurement uncertainty/error.

A GA for thermal design and optimization of a CHE was described. It was concluded that the GA has a strong ability for auto-search and combined optimization in the thermal design of heat exchangers without any trial-and-error process, and different objective functions (e.g., minimum volume, and minimum cost) may be considered.

NOMENCLATURE

A	coefficient	[kg/ms]
A	function	[-]
A	coefficient	
B	coefficient	
C	convective flux	
c_p	specific heat	[J/kgK]
D	diffusion coefficient	[kg/ms]
D	diameter	[m]
f	friction factor	
f_o	friction factor smooth surface	
k	turbulent kinetic energy	[m²/s]
NTU	number of transfer units	
Nu	Nusselt number	
Nu_o	Nusselt number smooth surface	
n	normal vector	
P	pitch	[m]
P	pressure	[Pa]
Pr	Prandtl number	

R	residual	
S	source term, pitch	
S_C	constant part of S	
S_P	linear coefficient of S	
t	time	[s]
u_i	velocity vector	[m/s]
u	velocity in x-direction	[m]
U_j	velocity vector	
u_i	velocity vector	[m/s]
v	velocity in y-direction	[m]
W	width	[m]
X,Y, Z	Cartesian coordinates	[m]
x_j	Cartesian coordinates	[m]

Greek symbols

ε	dissipation rate	$[m^2/s^3]$
ε	thermal efficiency	
φ	arbitrary variable	
Γ	diffusion coefficient	$[m^2/s]$
γ	exponent	
ρ	density	$[kg/m^3]$
ω	turbulence frequency	

Subscripts

E, W, N, S	east, west, north, and south grid point, respectively
e, w, n, s	east, west, north, and south face, respectively
f	face
NB	neighbor
P	arbitrary or general grid point

Abbreviations

ANN	artificial neural network
ASM	algebraic stress model
B	boundary
BEM	boundary element method
BPHE	brazed plate heat exchanger
CAD	computer aided design
CC	cross corrugated
CFD	computational fluid dynamics
CHE	compact heat exchanger
CI	computational intelligence
CVFEM	control volume finite element method
CW	cross wavy
DES	detached eddy simulation
DNS	direct numerical simulation
EASM	explicit algebraic stress model
FDM	finite difference method

FEM	finite element method
FL	fuzzy logic
FTHE	fin-and-tube heat exchanger
GA	genetic algorithm
GGDH	generalized gradient diffusion hypothesis
HVAC&R	heating, ventilation, air conditioning, and refrigeration
LES	large eddy simulation
PFCHE	plate-and-fin compact heat exchanger
PHE	plate heat exchanger
PISO	pressure implicit split operator
QUICK	quadratic upstream interpolation convective kinetics
RANS	Reynolds-averaged Navier–Stokes
RSM	Reynolds stress model
SA	simulated annealing
SGS	subgrid scale
SIMPLE	semi-implicit method for pressure-linked equations
SIMPLEC	semi-implicit method for pressure-linked equations–consistent
SIMPLER	semi-implicit method for pressure-linked equations–revised
SIMPLEX	semi-implicit method for pressure-linked equations–extended
SST	shear stress transport
TDMA	tridiagonal matrix algorithm
V2F	specific turbulence model
WET	wealth equals earnings × time

ACKNOWLEDGMENTS

The reported work and research results are based on projects that were financially supported by the Swedish Energy Agency, the Swedish Scientific Council, and industries such as Volvo Aero Corporation (now GKN Aerospace Engine Systems), Siemens Industrial Turbines, Alfa Laval, Nilcon Engineering, and others. Several doctoral students, post docs, and visiting researchers contributed to the various projects related to this chapter.

REFERENCES

1. R.K. Shah and D. Sekulic, *Fundamentals of Heat Exchanger Design*, Wiley, New York, 2003.
2. B. Sunden, *Heat Transfer and Heat Exchangers, Kirk-Othmer Encyclopedia in Chemical Technology*, Wiley, New York, 2007.
3. J.E. Hesselgreaves, *Compact Heat Exchangers—Selection, Design and Operation*, Pergamon, Amsterdam, 2001.
4. F. Coletti, *Heat Exchanger Design Handbook*, Begell House, USA.
5. B. Sunden, *Introduction to Heat Transfer*, WIT Press, Southampton, UK, 2012.
6. H.K. Versteeg and W. Malalasekera, *An Introduction to Computational Fluid Dynamics, the Finite Volume Method*, 2nd Ed., Pearson-Prentice Hall, New York, 2007.

7. J.H. Ferziger and M. Peric, *Computational Methods for Fluid Dynamics*, Springer-Verlag, Berlin, 1996.

8. M. Mohanraj, S. Jayaraj, and C. Muraleedharan, Applications of Artificial Neural Networks for Thermal Analysis of Heat Exchangers—A Review, *International Journal of Thermal Sciences*, Vol. 90, 150–172, 2015.

9. D. Geb, F. Zhou, G. DeMoulin, and I. Catton, Genetic Algorithm Optimization of a Finned-Tube Heat Exchanger Modeled with Volume-Averaging Theory, *ASME Journal of Heat Transfer*, Vol. 135, p. 082602, 2013.

10. B. Sunden and L. Wang, Relevance of Heat Transfer and Heat Exchangers for Greenhouse Gas Emissions, in *Advances in Heat Transfer Engineering*, eds. B. Sunden and J. Vilemas, pp. 1001–1012, Begell House, New York, 2003.

11. B. Sunden and L. Wang, Relevance of Heat Transfer and Heat Exchangers for Development of Sustainable Energy Systems, *AJTEC 203, TED-AJ03-603*, CD-Rom Proceedings, Honolulu, Hawaii, USA, 2003.

12. B. Sunden, High Temperature Heat Exchangers, *CHE2005-29*, in *5th International Conference Enhanced, Compact and Ultra-Compact Heat Exchangers: Science, Engineering and Technology*, eds. R.K. Shah, M. Ishizuka, T.M. Rudy, and V.W. Wadekar, pp. 226–238, CD-Rom Proceedings, Whistler, Canada, 2005.

13. L. Wang and B. Sunden, Design Methodology for Multistream Plate-Fin Heat Exchangers in Heat Exchanger Networks, *Heat Transfer Engineering*, Vol. 22, pp. 3–11, 2001.

14. S. Kakac, H. Liu, and A. Pramuanjaroenkij, *Heat Exchangers—Selection, Rating and Thermal Design*, CRC Press, London, 2012.

15. F.P. Incropera, D.P. DeWitt, T. Bergman, and A. Lavine, *Introduction to Heat Transfer*, 5th Ed., Wiley, New York, 2007.

16. W.M. Kays and A.L. London, *Compact Heat Exchangers*, 3rd Ed., McGraw-Hill, New York, 1984.

17. R.L. Webb and N.H. Kim, *Principles of Enhanced Heat Transfer*, 2nd Ed., Taylor and Francis, New York, 2005.

18. R.K. Shah, M.R. Heikal, B. Thonon, and P. Tochon, Progress in Numerical Analysis of Compact Heat Exchanger Surfaces, *Advances in Heat Transfer*, Vol. 34, pp. 363–443, 2001.

19. S.V. Patankar and D.B. Spalding, A Calculation Procedure for Transient and Steady State Behavior of Shell-and-Tube Heat exchangers, in *Heat Exchanger Design and Theory Source Book*, eds. N. Afgan and E.U. Schlunder, pp. 155–176, McGraw-Hill, New York, 1974.

20. M. Prithiviraj and M.J. Andrews, Three-Dimensional Numerical Simulation of Shell-and-Tube Heat Exchangers, Part I: Foundation and Fluid Mechanics, *Numerical Heat Transfer*, Vol. 33, no. Part A, pp. 799–816, 1998.

21. M. Prithiviraj and M.J. Andrews, Three-Dimensional Numerical Simulation of Shell-and-Tube Heat Exchangers, Part II: Heat Transfer, *Numerical Heat Transfer*, Vol. 33, no. Part A, pp. 817–828, 1998.

22. Y.L. He, W.Q. Tao, B. Deng, X. Li, and Y. Wu, Numerical Simulation and Experimental Study of Flow and Heat Transfer Characteristics of Shell Side Fluid in Shell and Tube Heat Exchangers, *CHE2005-05*, in *Enhanced, Compact and Ultra Compact Heat Exchangers: Science, Engineering and Technology*, eds. R.K. Shah, M. Ishizuka, T.M. Rudy, and V.W. Wadekar, pp. 29–42, ECI, Hoboken, NJ, 2005.

23. B. Sunden, Computational Fluid Dynamics in Research and Design of Heat Exchangers, *Heat Transfer Engineering*, Vol. 28, no. 11, pp. 898–910, 2007.

24. M.M.A. Bhutta, N. Hayat, M.H. Bashir, A.R. Khan, K.N. Ahmad, and S. Khan, CFD Applications in Various Heat Exchangers design: A Review, *Applied Thermal Engineering*, Vol. 32, pp. 1–12, 2012.

25. S.V. Patankar, *Numerical Heat Transfer and Fluid Flow*, McGraw-Hill, New York, 1980.

26. D.A. Andersson, J.C. Tannehill, and R.H. Pletcher, *Computational Fluid Mechanics and Heat Transfer*, 2nd Ed., Taylor and Francis, New York, USA, 1997.

27. G.D. Smith, *Numerical Solution of Partial Differential Equations*, Oxford University Press, London, 1978.

28. J.N. Reddy and D.K. Gartling, *The Finite Element Method in Heat Transfer and Fluid Dynamics*, CRC Press, Boca Raton, FL, 2010.

29. R.W, Lewis, K. Morgan, H.R. Thomas, and K.N. Seetharamu, *The Finite Element Method in Heat Transfer Analysis*, Wiley, Oxford, UK, 1996.

30. M.S. Kandelousi and D.D. Ganji, *Hydrothermal Analysis in Engineering Using Control Volume Finite Element Method*, Academic Press, Oxford, 2015.

31. L.C. Wrobel, *Boundary Element Method—Volume 1 Applications Thermo-Fluids and Acoustics*, Wiley, Oxford, UK, 2002.

32. C.M. Rhie and W.L. Chow, Numerical Study of the Turbulent Flow Past an Airfoil with Trailing Edge Separation, *AIAA Journal*, Vol. 21, pp. 1525–1532, 1983.

33. D.S. Jang, R. Jetli, and S. Acharya, Comparison of the PISO, SIMPLER and SIMPLEC Algorithms for Treatment of the Pressure Velocity Coupling in Steady Flow Problems, *Numerical Heat Transfer*, Vol. 10, no. 3, 209–228, 1986.

34. R.I. Issa, Solution of the Implicitly Discretized Fluid Flow Equations by Operator-Splitting, *Journal of Computational Physics*, Vol. 62, pp. 40–65, 1986.

35. D. McBride, N. Croftand, and M. Cross, Combined Vertex-Based-Cell-Centred Finite Volume Method for Flow in Complex geometries, *Third International Conference on CFD in the Minerals and Process Industries*, pp. 351–1356, CSIRO, Melbourne, Australia, 2003.

36. B. Farhanieh, L. Davidson, and B. Sunden, Employment of the Second-Moment Closure for Calculation of Recirculating Flows in Complex Geometries with Collocated Variable Arrangement, *International Journal of Numerical Methods in Fluids*, Vol. 16, pp. 525–554, 1993.

37. S. Pope, *Turbulent Flows*, Cambridge University Press, Cambridge, UK, 2000.

38. D.C. Wilcox, *Turbulence Modeling for CFD*, 2nd Ed., DCW Industries, La Canada, CA, 2002.

39. P.A. Durbin and T.I.-P. Shih, An Overview of Turbulence Modeling, in *Modeling and Simulation of Turbulent Heat Transfer*, eds. B. Sunden and M. Faghri, pp. 3–31, WIT Press, Southampton, UK, 2005.

40. P.R. Spalart and S.R. Allmaras, One-Equation Turbulence Model for Aerodynamic Flows, *AIAA Paper* -92-0439, AIAA collocated conferences, Nashville, TN, USA, 1992.

41. P.A. Durbin, Separated Flow Components with k-ε-v2 Model, *AIAA Journal*, Vol. 33, no. 4, pp. 659–664, 1995.

42. F.R. Menter, Zonal Two-Equation k-ω Models for Aerodynamic Flows, *AIAA Paper* 93–2906, AIAA collocated conferences, Orlando, FL, USA, 1993.

43. B.E. Launder, On the Computation of Convective Heat Transfer in Complex Turbulent Flows, *ASME Journal of Heat Transfer*, Vol. 110, pp. 1112–1128, 1988.

44. P.R. Spalart, W.-H. Jou, M. Stretlets, and S.R. Allmaras, Comments on the Feasibility of LES for Wings and the Hybrid RANS/LES Approach, *Advances in DNS/LES, Proceedings of the First AFOSR International Conference on DNS/LES*, 1997.

45. S.V. Patankar, C.H. Liu, and E.M. Sparrow, Fully Developed Flow and Heat Transfer in Ducts Having Streamwise-Periodic Variations of Cross-Sectional Area, *ASME Journal of Heat Transfer*, Vol. 99, pp. 180–186, 1977.

46. J. Klett, R. Hardy, E. Romine, C. Walls, and T. Burchell, High-Thermal-Conductivity, Mesophase-Pitch-Derived Carbon Foams: Effect of Precursor on Structure and Properties, *Carbon*, Vol. 38, pp. 953–973, 2000.

47. Q. Wang, X.H. Han, A. Sommers, Y. Park, C.T. Joen, and A. Jacobi, A Review on Application of Carbanaceous Materials and Carbon Matrix Composites for Heat Exchangers and Heat Sinks, *International Journal of Refrigeration*, Vol. 35, pp. 7–26, 2012.

48. A.G. Straatman, N.C. Gallego, Q. Yu, and B.E. Thomson, Characterization of Porous Carbon Foam as a Material for Compact Recuperators, *ASME Journal of Engineering for Gas Turbines and Power*, Vol. 129, pp. 326–330, 2007.

49. N.C. Gallego and J.W. Klett, Carbon Foams for Thermal Management, *Carbon*, Vol. 41, pp. 1461–1466, 2003.

50. K.C. Leong, L.W. Jin, H.Y. Li, and J.C. Chai, Forced Convection Air Cooling in Porous Graphite Foam for Thermal Management Applications, *Proceedings of the 11th Intersociety Conference on Thermal and Thermomechanical Phenomena in Electronic Systems*, pp. 57–64, 2008.

51. Y.R. Lin, J.H. Du, W. Wu, L. Chow, and W. Notardonato, Experimental Study on Heat Transfer and Pressure Drop of Recuperative Heat Exchangers using Carbon Foam, *ASME Journal of Heat Transfer*, Vol. 132, paper no 091902, 2010.

52. M. Sen and K.T. Yang, Applications of Artificial Neural Networks and Genetic Algorithms in Thermal Engineering, in *The CRC Handbook of Thermal Engineering*, ed. F. Kreith, pp. 620–661, CRC Press, Boca Raton, FL, 2000.

53. K.T. Yang and M. Sen, Artificial Neural Network-Based Dynamic Modeling Thermal Systems and their Control, in *Heat Transfer Science and Technology*, ed. Q. Bu-Xuan, Higher Education Press, Beijing, 2000.

54. G. Diaz, M. Sen, K.T. Yang, and R.T. McClain, Simulation of Heat Exchanger Performance by Artificial Neural Networks, *International Journal of HVAC&R Research* Vol. 5, pp. 195–208, 1999.

55. G. Diaz, M. Sen, K.T. Yang, and R.T. McClain, Dynamic Prediction and Control of Heat Exchangers Using Artificial Neural Networks, *International Journal of Heat Mass Transfer,* Vol. 45, pp. 1671–1679, 2001.

56. G. Diaz, M. Sen, K.T. Yang, and R.T. McClain, Adaptive Neuro-Control of Heat Exchangers, *ASME Journal of Heat Transfer* Vol. 123, pp. 417–612, 2001.

57. G. Diaz, M. Sen, K.T. Yang, and R.T. McClain, Stabilization of Thermal Neuro-Controllers, *Applied Artificial Intelligence,* Vol. 18, pp. 447–466, 2004.

58. A. Pacheco-Vega, G. Diaz, M. Sen, K.T. Yang, and R.T. McClain, Neural Network Analysis of Fin-Tube Refrigerating Heat Exchanger with Limited Experimental Data, *International Journal of Heat Mass Transfer*, Vol. 44, pp. 763–770, 2000.

59. A. Pacheco-Vega, G. Diaz, M. Sen, K.T. Yang, and R.T. McClain, Heat Rate Predictions in Humid Air-Water Heat Exchangers Using Correlations and Neural Networks, *ASME Journal of Heat Transfer*, Vol. 123, pp. 348–354, 2001.

60. Y. Islamoglu, A New Approach for the Prediction of the Heat Transfer Rate of the Wire-on-Tube Type Heat Exchanger-Use of an Artificial Neural Network Model, *Applied Thermal Engineering*, Vol. 23, pp. 243–249, 2003.

61. Y. Islamoglu, A. Kurt, and C. Parmaksizoglu, Performance Prediction for Non-Adiabatic Capillary Tube Suction Line Heat Exchanger: An Artificial Neural Network Approach, *Energy Conversion and Management*, Vol. 46, pp. 223–232, 2005.

62. G.N. Xie, Q.W. Wang, M. Zeng, and L.Q. Luo, Heat Transfer Analysis for Shell-and-Tube Heat Exchangers with Experimental Data by Artificial Neural Networks Approach, *Applied Thermal Engineering*, Vol. 27, pp. 1096–1104, 2007.

63. Q.W. Wang, G.N. Xie, M. Zeng, and L.Q. Luo, Prediction of Heat Transfer Rates for Shell-and-Tube Heat Exchangers by Artificial Neural Network Approach, *Journal of Thermal Science,* Vol. 15, pp. 257–262, 2006.

64. H.M. Ertunc and M. Hosoz, Artificial Neural Network Analysis of a Refrigeration System with an Evaporative Condenser, *Applied Thermal Engineering*, Vol. 26, pp. 627–635, 2007.

65. M. Hosoz, H.M. Ertunc, and H. Bulgurcu, Performance Prediction of a Cooling Tower Using Artificial Neural Network, *Energy Conversion and Management*, Vol. 48, pp. 1349–1359, 2007.
66. K.S. Yigit and H.M. Ertunc, Prediction of the Air Temperature and Humidity at the Outlet of a Cooling Coil Using Neural Networks, *International Communications in Heat and Mass Transfer*, Vol. 33, pp. 898–907, 2006.
67. G.J. Zdaniuk, L.M. Chamra, and D.K. Walters, Correlating Heat Transfer and Friction in Helically-Finned Tubes Using Artificial Neural Networks, *International Journal of Heat and Mass Transfer*, Vol. 50, pp. 4713–4723, 2007.
68. Y. Islamoglu and A. Kurt, Heat Transfer Analysis Using ANNs with Experimental Data with Air Flow in Corrugated Channels, *International Journal of Heat and Mass Transfer*, Vol. 47, pp. 1361–1365, 2004.
69. A.J. Ghajar, L.M. Tam, and S.C. Tam, Improved Heat Transfer Correlation in Transition Region for a Circular Tube with Three Inlet Configurations Using Artificial Neural Networks, *Heat Transfer Engineering*, Vol. 25, no. 2, pp. 30–40, 2004.
70. S.S. Sablani, A. Kacimov, J. Perret, A.S. Mujumdar, and A. Campo, Non-Iterative Estimation of Heat Transfer Coefficients Using Artificial Neural Network Models, *International Journal of Heat and Mass Transfer*, Vol. 48, pp. 665–679, 2005.
71. S. Deng and Y. Hwang, Applying Neural Networks to the Solution of Forward and Inverse Heat Conduction Problems, *International Journal of Heat and Mass Transfer*, Vol. 49, pp. 4732–4750, 2006.
72. S. Deng and Y. Hwang, Solution of Inverse Heat Conduction Problems Using Kalman Filter-Enhanced Bayesian Back Propagation Neural Network Data fusion, *International Journal of Heat and Mass Transfer*, Vol. 50, pp. 2089–2100, 2007.
73. K. Ermis, A. Erek, and I. Dincer, Heat Transfer Analysis of Phase Change Process in a Finned-Tube Thermal Energy Storage System Using Artificial Neural Network, *International Journal of Heat and Mass Transfer*, Vol. 50, pp. 3163–3175, 2007.
74. B. Ayhan-Sarac, B. Karlık, T. Bali, and T. Ayhan, Neural Network Methodology for Heat Transfer Enhancement Data, *International Journal of Numerical Methods for Heat Fluid Flow*, Vol. 17, pp. 788–798, 2007.
75. G.N. Xie, L.H. Tang, B. Sunden, and Q.W. Wang, Artificial Neural Network Based Correlating Heat Transfer and Friction Factor of Fin-and-Tube Heat Exchanger with Large Number of Large-Diameter Tube Rows, *ASME HT2008-56241*, ASME Heat Transfer Conference, Jacksonville, FL, USA, 2008.
76. G. Xie, B. Sunden, Q.W. Wang, and L. Tang, Performance Prediction of Laminar and Turbulent Heat Transfer and Fluid Flow Heat Exchangers Having Large Tube Diameter and Large Tube Row by Artificial Neural Networks, *International Journal of Heat and Mass Transfer*, Vol. 52, no. 11–12, pp. 2484–2497, 2009.
77. A. Pacheco-Vega, M. Sen, K.T. Yang, and R.L. McClain, Genetic Algorithms-Based Predictions of Fin-Tube Heat Exchanger Performance, *Proceedings of 11th International Heat Transfer Conference*, Vol. 6, pp. 137–142, 1998.
78. M.C. Tayal, Y. Fu, and U.M. Diwekar, Optimal Design of Heat Exchangers: A Genetic Algorithm Framework, *Industry Engineering and Chemical Research*, Vol. 38, pp. 456–467, 1999.
79. R. Selbas, O. Kizilkan, and M. Reppich, A New Design Approach for Shell-and-Tube Heat Exchangers Using Genetic Algorithms from Economic Point of View, *Chemical Engineering and Processing*, Vol. 45, no. 4, pp. 268–275, 2006.
80. H.X. Liang, G.N. Xie, M. Zeng, Q.W. Wang, and Z.P. Feng, Application of Genetic Algorithm to Optimization Recuperator in Micro-Turbine, *The 2nd International Symposium on Thermal Science and Engineering*, October 23–25, Beijing, China, 2005.

81. H.X. Liang, G.N. Xie, M. Zeng, Q.W. Wang, and Z.P. Feng, Genetic Algorithm Optimization for Primary Surfaces Recuperator of Microturbine, *ASME Turbo Expo 2006, paper no GT2006-90366*, ASME IGTI International Gas Turbine Conference, Barcelona, Spain, 2006.

82. A. Pacheco-Vega, M. Sen, K.T. Yang, and R.L. McClain, Correlations of Fin-Tube Heat Exchanger Performance Data Using Genetic Algorithms Simulated Annealing and Interval Methods, *Proceedings of ASME the Heat Transfer Division,* Vol. 369-5, pp. 143–151, 2001.

83. A. Pacheco-Vega, M. Sen, and K.T. Yang, Simultaneous Determination of In- and Over-Tube Heat Transfer Correlations in Heat Exchangers by Global Regression, *International Journal of Heat and Mass Transfer,* Vol. 46, no. 6, pp. 1029–1040, 2003.

84. M. Mishra, P.K. Das, and S. Saranqi, Optimum Design of Crossflow Plate-Fin Heat Exchangers through Genetic Algorithm, *International Journal of Heat Exchangers,* Vol. 5, no. 2, pp. 379–401, 2004.

85. I. Ozkol and G. Komurgoz, Determination of the Optimum Geometry of the Heat Exchanger Body via a Genetic Algorithm, *Numerical Heat Transfer, Part A,* Vol. 48, pp. 283–296, 2005.

86. G.N. Xie and Q.W. Wang, Geometrical Optimization of Plate-Fin Heat Exchanger Using Genetic Algorithms, *Proceedings of the Chinese Society for Electrical Engineering,* Vol. 26, no. 7, pp. 53–57, 2006.

87. D.E. Goldberg, *Genetic Algorithms in Search, Optimization and Machine Learning,* Addison-Wesley Publishing Company, New York, USA, 1989.

88. Z. Michalewicz, *Genetic Algorithms + Data Structures = Evolution Programs,* Springer-Verlag, New York, 1999.

89. R.L. Haupt and S.E. Haupt, *Practical Genetic Algorithms,* 2nd Ed., Wiley, Oxford, UK, 2004.

90. D.L. Carroll, Chemical Laser Modeling with Genetic Algorithms, *AIAA Journal,* Vol. 34, no. 2, pp. 338–346, 1996.

91. D.L. Carroll, Genetic Algorithms and Optimizing Chemical Oxygen-Iodine Lasers, in *Developments in Theoretical and Applied Mechanics,* eds. H. Wilson, R. Batra, C. Bert, A. Davis, B. Schapery, B. Stewart, and F. Swinson, pp. 411–424, Vol. XVIII, School of Engineering, University of Alabama, Huntsville, Alabama, USA, 1996.

92. Z. Michalewicz, K. Deb, M. Schmidt, and T. Stidsen, Test-Case Generator for Constrained Parameter Optimization Techniques, *IEEE Transactions on Evolutionary Computation,* Vol. 4, no. 3, pp. 197–215, 2000.

93. V.S. Gullapalli and B. Sunden, Generalized Performance Analysis of Compact Brazed Plate Heat Exchangers, *21st National and 10th ISHMT-ASME Heat and Mass Transfer Conference,* IIT Madras, India, 2011.

94. V.S. Gullapalli and B. Sunden, CFD Simulation of Heat Transfer and Pressure Drop in Compact Brazed Plate Heat Exchangers, *Heat Transfer Engineering,* Vol. 35, no. 4, 358–366, 2014.

95. S. Etemad and B. Sunden, Hydraulic and Thermal Simulations of a Cross-Corrugated Plate Heat Exchanger Unitary Cell, *Heat Transfer Engineering,* Vol. 37, no. 5, 475–486, 2016.

96. B. Sunden, On Partially Corrugated Ducts in Heat Exchangers, *ASME/IMECE 2002 CD-ROM Proceedings, IMECE2002-39651,* 2002.

97. W.M. Lin, B. Sundén, and J.L. Yuan, A Performance Analysis of Porous Graphite Foam Heat Exchangers in Vehicles, *Applied Thermal Engineering,* Vol. 50, pp. 1201–1210, 2013.

98. W. Lin, J. Yuan, and B. Sunden, Performance Analysis of Aluminum and Graphite Foam Heat Exchangers Under Countercurrent Arrangement, *Heat Transfer Engineering,* Vol. 35, no. 6–8, pp. 730–737, 2014.

99. W. Lin, G. Xie, B. Sunden, and Q.W. Wang, Flow and Thermal Performance of Graphite Foam Dimpled Fin Heat Exchangers, *Proceedings of 15th International Heat Transfer Conference, paper no IHTC15-8536*, 2014.

100. G. Xie, B. Sunden, Q.W. Wang, and L. Tang, Performance Prediction of Laminar and Turbulent Heat Transfer and Fluid Flow Heat Exchangers Having Large Tube Diameter and Large Tube Row by Artificial Neural Networks, *International Journal of Heat and Mass Transfer*, Vol. 52, no 11–12, pp. 2484–2497, 2009.

101. Q. Wang, G. Xie, and B. Sunden, Design Optimization and Performance Prediction of Compact Heat Exchangers, in *Emerging Topics in Heat Transfer-Enhancement and Heat Exchangers*, eds. Q.W. Wang, Y. Chen, and B. Sunden, pp. 301–319, WIT Press, Southampton, UK, 2014.

102. G. Xie, Q. Wang, and B. Sunden, Application of a Genetic Algorithm for Thermal Design of Fin-and-Tube Heat Exchangers, *Heat Transfer Engineering*, Vol. 29, no. 7, pp. 597–607, 2008.

103. G. Xie, B. Sunden, and Q. Wang, Optimization of Compact Heat Exchangers by a Genetic Algorithm, *Applied Thermal Engineering*, Vol. 28, no. 8–9, pp. 895–906, 2008.

104. G. Xie, Q.W. Wang, and B. Sunden, Parameter Study and Multiple Correlations on Air-Side Heat Transfer and Friction Characteristics of Fin-and-Tube Heat Exchangers with Large Number of Large Diameter Tube Rows, *Applied Thermal Engineering*, Vol. 29, pp. 1–16, 2009.

3 Utilization of Numerical Methods and Experiments for the Design and Tests of Gasketed Plate Heat Exchangers

Selin Aradag, Sadik Kakac, and Ece Özkaya

CONTENTS

ABSTRACT: This chapter mainly focuses on numerical studies performed to analyze a plate heat exchanger composed of specific plates and the experimental validation of the numerical study. Different turbulence models, mainly k-ε, RNG k-ε, EARSM k-ε, k-ω, and SST k-ω, are used and compared in computational fluid dynamics analysis. First, the numerical methodology is validated with the help of experiments. The effects of several geometrical properties on the performance of the exchangers are investigated numerically with the help of the validated methodology. These geometrical properties include channel height, wave amplitude, and distribution channels. New plate designs with better thermal and hydraulic performance are obtained with the utilization of experimentally validated computational fluid dynamics (CFD) techniques.

3.1   INTRODUCTION

A heat exchanger is a device that allows the exchange of heat between two fluids without their mixing. Heat exchangers have a wide range of application areas including heating, cooling, and ventilation [1–3]. They are classified according to heat generation/regeneration capabilities, transfer time, geometrical properties, the mechanism of heat transfer, and flow arrangement. According to their classification with respect to geometrical properties, plate heat exchangers are important and widely used.

Although shell-and-tube heat exchangers are the most well-known type, Chevron-type plate heat exchangers have been enhanced in recent years with the help of advanced materials and production technologies, and they currently have a wide range of applicability. Since their first production in 1930s, plate heat exchangers have been used for many years in various fields as reported by Kakac et al. [4]. They are more practical and effective when compared to most heat exchanger types; therefore, the utilization of gasketed plate heat exchangers in industrial applications has increased drastically in recent years. Even though much information has been gained about plate heat exchangers, today it cannot be said that there are any generalized formulas or equations to characterize their thermal-hydraulic performance due to their complex geometrical configurations.

Gasketed plate heat exchangers have several parts: the plate pack, which is the part that participates in heat transfer; gaskets; and inlet and outlet ports are some of the important, as shown in Figure 3.1.

There are several advantages of gasketed plate heat exchangers when compared to shell-and-tube type heat exchangers. For example, geometrical features (mainly the indents on the plates) help turbulence develop and increase heat transfer as well as the total heat transfer area. Gasketed plate heat exchangers are compact.

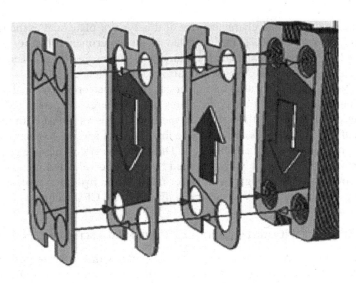

FIGURE 3.1 Plate arrangement for gasketed plate heat exchangers. (Adapted from Kakac, S., et al., *Eng. Mach.*, 54, 44–68.)

When compared to shell-and-tube heat exchangers of the same total heat transfer area, they can be as much as 30% smaller in terms of mass and 20% smaller in terms of volume. They have a thermal recovery ratio of around 90%, which makes them suitable for thermal recovery applications—even for low temperature differences. Their gaskets are convenient to clean and maintain, and they are very easy to construct. The number of plates can be readily changed to design a new heat exchanger package. Also, they are easy to control. It is possible to work with two different fluid types. Thermal losses are minor because only the end plates are open to the atmosphere. However, the gasket type and the maximum working temperature and pressure are its limitations.

The lengthy experimental procedure for plate heat exchanger design requires extensive time and effort. From such experiments, it is necessary to develop specific correlations of heat transfer and pressure drop for the particular geometries of the plate heat exchangers used. Constraints such as the dimensions of the heat exchanger, types of connections, and the type and mass flow rate of the fluid utilized that profoundly affect performance. CFD is a relatively easier method compared to experiments provided that the methodology and the results are verified with the experimental findings. Local values of temperature, pressure, and velocity can also be observed easily with the help of computational results.

3.1.1 Objective of the Study

First, a general overview of numerical studies related to gasketed plate heat exchangers and their experimental validation is presented. The TOBB University of Economics and Technology Heat Exchanger Design Laboratory is capable of testing the thermal and hydraulic characteristics of this type of plate heat exchangers with the help of experiments [6,7]. The overview of the experimental facility is given in Section 3.2.1. As well as experiments, numerical studies and CFD-aided design of plates for plate type heat exchangers are also major challenges.

This chapter mainly focuses on numerical studies performed to analyze a plate heat exchanger composed of specific plates and on the experimental validation of the numerical study. Once the numerical methodology is validated with the help of experiments, the next step is to investigate the effects of several geometrical properties on the performance of the exchangers. Therefore, the effects of plate geometry on heat exchanger performance is investigated by using the validated numerical methodology, and new plate designs with better thermal and hydraulic performance are obtained with the utilization of experimentally validated CFD techniques.

3.1.2 REVIEW OF NUMERICAL STUDIES OF PLATE HEAT EXCHANGERS IN LITERATURE

In the literature, there are many experimental studies that analyze gasketed plate heat exchangers thermally and hydraulically. Also, in recent years, CFD has become another working field for evaluating hydraulic and thermal performance of plate heat exchangers, as stated by Wang et al. [3]. A study about the laminar to transitional region is investigated by Blomerius and Mitra [8] for a Reynolds number range of 600–2000. In this study, dimensionless wave length and dimensionless channel height are optimized for the best ratio of heat transfer to pressure drop (performance evaluation criterion) by two-dimensional (2D) CFD analyses. The optimum values according to the performance evaluation criterion used are 12 and 2, respectively. The flow structure and heat transfer are investigated for three dimensional (3D) channels, constructed according to optimum values from the 2D analysis. In addition to dimensionless wave length and dimensionless channel height, corrugation angle (φ) is an important parameter for 3D channel geometry, which is chosen as 45° and 90°. It is seen that CFD analyses of φ of 90° are more accurate than 45°. Oscillations occur in the flow when a critical Reynolds number is reached. Two-dimensional analyses estimate a critical Reynolds number lower than 3D analyses. It is also noted that the oscillations increase the heat transfer. It is shown that the 45°corrugation angle of 3D analyses gives the best performance.

Ismail and Velraj [9] investigated the friction factor (f) value and Colburn factor (j) on offset and wavy fins. In this study, air is used as the working fluid, and the CFD methodology is validated with available literature and experimental results of f and j correlations. CFD results of offset fins are compared to in-house experimental results as well as to correlations from literature, where CFD results of 18 different wavy fins are compared to limited correlations available in the literature. For offset fins, it is noted that j estimation is in good agreement, while f correlation needs a multiplication factor of 1.6 for design considerations. For wavy fins, two different correlations are developed for laminar (Reynolds range of 100–800) and turbulent (Reynolds range of 1000–15,000) flow regimes.

Pelletier et al. [10] investigated whether simulation of a brazed plate heat exchanger (BPHE) for determining heat transfer characteristics can be conducted by using a FLUENT CFD simulation program or not. The k-ω SST turbulence model for transitional flow is used where the Reynolds number is about 3500. The working fluid is water, and its properties are constant. Constant heat flow and constant wall temperature conditions are used as boundary conditions in CFD analyses. CFD results are compared with the results from experiments conducted for plates with the same geometric features. It is observed that the simulation shows a deviation of 45% from the experimental values for a constant wall temperature boundary condition and 58% for a constant heat flux boundary condition; 32 different arrangements of the effect of flow arrangement on pressure drop are studied by Miura et al. [11]. A correlation for total pressure drop is developed and generalized with arrangement parameters: mean channel flow rate and number of passes. Also, a friction factor correlation is developed for a Reynolds range of 10–870. Turbulence model study concluded that a k-ω SST turbulence model predicts pressure drop value more accurately than a k-ε turbulence model. CFD results for a k-ω SST turbulence model are 10% to 13% lower than experimental results. The study also shows that a higher mass flow rate leads to a maldistribution of flow between the first and second channels.

Experimental and numerical studies for a chevron plate heat exchanger were carried out by Jain et al. [12]. The chevron plate angle used in experiments and numerical studies is 60°. The flow arrangement is single plate, counter flow. A numerical model is developed and validated with experimental data and empirical correlations from literature. The experimental results are obtained in a Reynolds range of 400–1300 and Prandtl range of 4.4–6.3. In the study, one channel is used for the cold side and two halves of a channel for the hot side. The cold channel has corrugated walls on both sides, whereas hot channels have one corrugated and one flat side. The flat sides of the hot channels are used for periodic boundary conditions, and the CFD model represents an infinite pack. A realizable k-ε turbulence model is used, and the numerical friction factor is underpredicted by 10% on average, the numerical Nusselt number is underpredicted by 13% on average.

O'Halloran and Jokar [13] studied a BPHE experimentally and numerically. Using commercial code, FLUENT, numerical analyses are performed for three BPHEs that have different chevron angles of 60°/60°, 27°/60°, and 27°/27°. Various temperature and velocity values are utilized as boundary conditions. The k-ω SST turbulence model is used. In this study, outlet temperature, pressure drop, and heat transfer are compared for three different chevron-type plates. They showed that improved CFD models can be used for determining heat transfer and fluid flow characteristics. Kanaris et al. [14] performed an experimental and numerical study on heat transfer enhancement and fluid motion inside channels. CFD results are validated with the pressure drop and temperature difference values. Fluid temperature profile inside the corrugated channel and temperature difference are obtained from an infrared thermography camera and compared with the experimental results. It is found that plate heat exchanger simulations with a commercial CFD code are an effective tool for predicting flow characteristics, heat transfer, and pressure drops. Han et al. [15] investigated thermal and hydraulic characteristics of a chevron-type plate heat exchanger that is made up of sinusoidal corrugation shaped plates. A k-ω SST turbulence model is used in simulations. In order to validate simulation results, the results are compared with correlations that are derived from an experimental data ensemble. It is found that CFD results are in good agreement with experimental results and maximum deviation is ± 20%. Patil et al. [16] investigated the usage of CFD application to optimize design parameters for a milk pasteurizer plant. Temperature distribution, flow configuration, and material thermal conductivity of a plate heat exchanger that is widely used in daily in a milk pasteurizer plant is studied experimentally and numerically. Simulation results deviate from experimental results by approximately 2%. CFD simulations show that heat transfer capabilities of plate heat exchangers enhance as thermal conductivity of plate material increases.

3.2 METHODOLOGY

First, the experimental methodology and set-up used to test the thermal and hydraulic performances of gasketed plate heat exchangers are explained briefly in Section 3.2.1. The experiments and the experimental findings are explained in detail by several publications of our group [6,7]; therefore, the concentration of this work is on the numerical studies performed. The numerical methodology for the CFD simulations is explained in detail in Section 3.2.2. The results of the computations are validated by the use of the experimentally obtained results.

3.2.1 EXPERIMENTAL METHODOLOGY

The experimental set-up uses tap water as the working fluid to carry out experiments with a wide range of Reynolds numbers and Prandtl number values. Water flows in two different circuits— the hot and cold water circuits. Water is heated with the help of resistance heaters. In the cold water circuit, cold water is pumped into the tested heat exchanger from the cold water tank (Figure 3.2). During the experiments, discharge water is collected at a waste water tank.

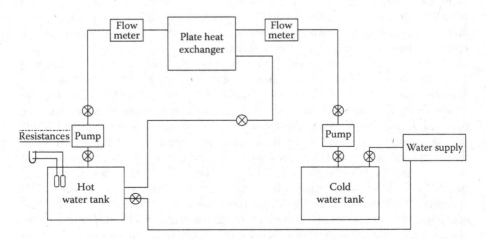

FIGURE 3.2 Schematic view of the experimental set-up. (Adapted from Akturk, F., et al., *J. Therm. Sci. Technol.*, 35(1), 43–52, 2015.)

Both of the circuits are equipped with electromagnetic flow meters to measure the flow rate. Thermocouples are used for temperature measurements; pressure transmitters are used to measure the inlet and outlet pressure differences. After the stabilization of the system, temperature values, pressure differences, and volumetric flow rates for each loop are recorded [7]. A gasketed plate heat exchanger that is composed of commercial chevron plates is utilized in the experimental analysis. Table 3.1 shows the geometrical parameters associated with the gasketed plates; these dimensions are also shown on the plate geometry in Figure 3.3. The plate heat exchanger is configured in a single pass, cross flow, and U-type flow arrangement.

Experiments are performed in our laboratory [6,7] for a wide range of Reynolds numbers (300–5000) in order to obtain heat transfer and pressure drop values. The correlations for Nusselt number and friction factor are derived from the experiments in the form of Equations 3.1 and 3.2 for a chevron angle of 30°, with an uncertainty of ±12.4% and ±11.8%, respectively [6].

$$\mathrm{Nu} = 0.32867 \mathrm{Re}^{0.68} \mathrm{Pr}^{1/3} \left(\frac{\mu}{\mu_w} \right)^{0.14} \quad 300 < \mathrm{Re} < 500 \quad\quad (3.1)$$

$$f = 259.9 \mathrm{Re}^{-0.9227} + 1.246 \quad\quad 300 < \mathrm{Re}\, 500 \quad\quad (3.2)$$

Different correlations are developed for various plate types with different numbers of plates, and the results are presented in detail in our group's publications [6,7]. Figure 3.4 shows representative results for Nusselt numbers for two different plates. Plate 1 is the plate tested numerically in this work. Figure 3.5 shows the comparison of the results for Nusselt numbers with two other results from the literature. As shown in Figure 3.5, every plate is unique; the characteristics must be developed

TABLE 3.1
Chevron Plate Properties

Plate Parameters	Parameter Definition	Plate Property
β (°)	Chevron angle	30
D_p (m)	Port diameter	0.035
L_w (m)	Port to port plate width	0.109
L_v (m)	Chevron area length	0.37
L_p (m)	Port to port length	0.335
b (m)	Corrugation depth	2.76
t (m)	Plate thickness	0.45
A_l (m)	Effective corrugated area	0.035
A_{lp} (m)	Projected surface area	0.03
ϕ	Enlargement factor	1.17
D_e (m)	Equivalent diameter	0.0055
D_h (m)	Hydraulic diameter	0.0047
k_w (W/m·K)	Thermal conductivity of the wall	16.2

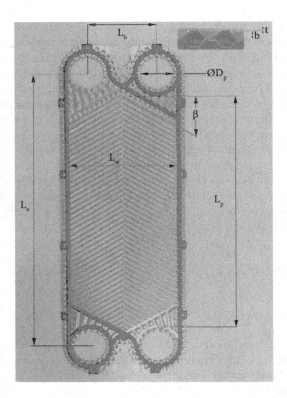

FIGURE 3.3 Dimensions of the plate tested experimentally and numerically. (Adapted from Gulenoglu, C., et al., *Int. J. Therm. Sci.*, 75, 249–256, 2014.)

individually using experimental or numerical techniques, and the heat exchanger designs must be performed accordingly.

3.2.2 METHODOLOGY FOR THE NUMERICAL STUDIES AND VALIDATION OF THE RESULTS WITH EXPERIMENTS

3.2.2.1 Preparation of the Geometry and Mesh Generation

The commercial plates that are experimentally tested are scanned using optical scanning methods to be able to generate the 3D geometries that will be used in the computational studies. The solid model is generated with the help of the point cloud obtained from scanning. The solid model is shown in Figure 3.6a, whereas Figure 3.6b shows the wet surface that is used for the computations (where the part inside the gasket is removed). Figure 3.6c shows the computational domain used for the simulations, where both the hot and cold fluid surfaces are connected to each other. The distances between the ports are taken directly from the experimental set-up and added to the domain. The computational domain that also involves the ports for cold and hot fluid inlets and outlets is shown in Figure 3.6d. The mesh for the computational domain is shown in Figure 3.6e.

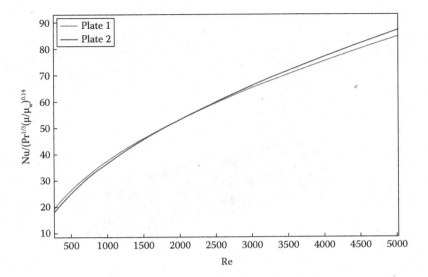

FIGURE 3.4 Experimental Nusselt number correlations. (Adapted from Gulenoglu, C., et al., *Int. J. Therm. Sci.*, 75, 249–256, 2014.)

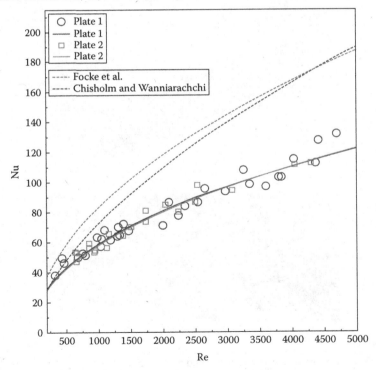

FIGURE 3.5 Comparison of Nusselt numbers obtained from experiments with literature. (Adapted from Gulenoglu, C., et al., *Int. J. Therm. Sci.*, 75, 249–256, 2014.)

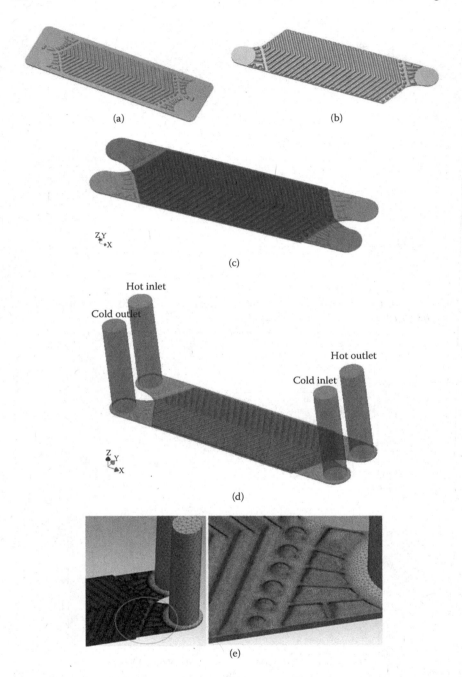

FIGURE 3.6 Computational geometry: (a) solid model for the plate, (b) solid model used for the computations, (c) computational fluid dynamics (CFD) domain, (d) final CFD domain including the inlet and outlet ports for cold and hot fluids, and (e) mesh for the computational domain.

3.2.2.2 CFD Methodology

In the CFD model, the physical model is simplified using the following assumptions in order to reduce the calculation cost. Both cold and hot channels are free from fouling, heat transfer and flow characteristics are in steady-state condition, and no slip wall condition is assumed for all walls. The geometrical parameters associated with the plate are shown in Figure 3.7.

Boundary conditions for the CFD model are listed in Table 3.2. Mass flow rates of inlet boundary conditions are chosen using the experimental input.

Turbulence modeling is important for the CFD analysis. Different turbulence models represent different physical phenomena; therefore, turbulence model study is a requirement for further research to validate the model of each problem. Zhang and Che [17] investigated eight different turbulence models for a cross-corrugated plate in a unitary cell. In the study by Zhang and Che [17], five different turbulence models are investigated that are suitable for this study. The turbulence models investigated are k-ε and k-ω standard turbulence models and the developed versions of these two models: RNG k-ε, EARSM k-ε, and SST k-ω.

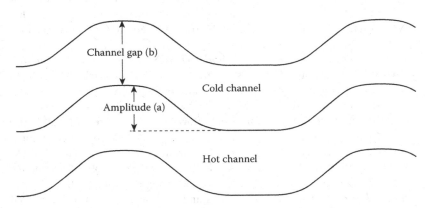

FIGURE 3.7   Schematic illustration for corrugation profiles of plates.

TABLE 3.2
Boundary Conditions of CFD Analyses

Surface	Boundary Condition	Value
Hot side: Inlet	Mass flow rate	0.03–0.13 kg/s
	Constant temperature	65°C
Hot side: Outlet	Static pressure	0 atm
Cold side: Inlet	Mass flow rate	0.03–0.13 kg/s
	Constant temperature	20°C
Cold side: Outlet	Static pressure	0 atm
Wall	No slip wall	Smooth wall
Plate	No slip wall	Smooth wall
	Thin material	0.45 mm steel

The turbulence methods used in the study are briefly discussed here:

Standard k-ε turbulence model. The k-ε model is one of the most well-known turbulence models which has been implemented in commercial CFD codes. It has proven to be stable and reliable. The k-ε model makes good predictions in terms of accuracy and robustness for general engineering flow problems but there are applications for which k-ε model may not be suitable. Among these are flows with boundary layer separation and sudden changes in the mean strain rate, flows in rotating fluids, and flows over curved surfaces [18].

Renormalization group (RNG) k-ε turbulence model. Compared to standard k-ε model, the RNG model has an additional term in its ε equation that improves the accuracy for strained flows. The RNG-based k-ε turbulence model is derived from the Navier–Stokes equations, using a mathematical technique called "renormalization group" methods. The RNG theory accounts for low-Reynolds number effects while the standard k-ε model is a high-Reynolds number model. The RNG k-ε model is more accurate and reliable for a wider class of flows than the standard k-ε model. Additional terms and functions in the transport equations are derived for k and ε [18].

Explicit algebraic Reynolds stress model (EARSM) k-ε turbulence model. EARSM is an extension of the standard two-equation models. It is derived from the Reynolds stress transport equations and gives a nonlinear relation between the Reynolds stresses, the mean strain rate, and vorticity tensors. Due to higher order terms, many flow phenomena are included in the model without solving transport equations. The EARSM model is complex because of the high amount of algebraic parameters to decide. The EARSM is an extension of the current k-ε model to capture effects of secondary flows, flows with streamline curvature, and system rotation [18].

Standard k-ω turbulence model. The standard k-ω model is an empirical example based on transport equations for the turbulence kinetic energy (k) and the turbulent frequency (ω). One of the advantages of the k-ω model is the near-wall treatment for low-Reynolds number computations. The model does not involve the complex nonlinear damping functions required for the k-ε model; therefore, it is more accurate and more robust. The k-ω models assume that the turbulence viscosity is a multiplication of density by the ratio of the turbulence kinetic energy and the turbulent frequency [18].

Shear stress transport (SST) k-ω turbulence model. The k-ω based SST turbulence model has become a very popular two-equation model. The amount of flow separation under adverse pressure gradients by the inclusion of transport effects into the formulation of the eddy viscosity is more accurate. The SST k-ω model can be used as a low-Reynolds number turbulence model without any extra near-wall treatments or functions. The SST model is recommended for high accuracy boundary layer simulations. The SST model switches into a k-ε model in the free-stream flow, which avoids the sensitivity problem of k-ω [18].

TABLE 3.3

Computational Resources

	Processor	RAM	Operating System	Number of Processors	Computational Time
Work station	Intel (R) Xeon (R) CPU E5-1620 3.6 GHz	64 GB	Windows 7 64-bit	8	48 hours/ processor
Cluster			Linux	108	14 hours/ 12 processors

The continuity, momentum, and energy equations are solved using a high order advection scheme, whereas the turbulence equations are solved using a first order upwind scheme. The conservation equations for mass, momentum, and energy are given in Equations 3.3–3.5:

$$\frac{\partial \rho}{\partial t} + \frac{\partial}{\partial x_j}\left(\rho U_j\right) = 0 \tag{3.3}$$

$$\frac{\partial}{\partial t}\left(\rho U_i\right) + \frac{\partial}{\partial x_j}\left(\rho U_j U_i\right) = -\frac{\partial P}{\partial x_i} + \frac{\partial}{\partial x_j}\left(\mu_{\text{eff}}\left(\frac{\partial U_i}{\partial x_j} + \frac{\partial U_j}{\partial x_i}\right)\right) \tag{3.4}$$

$$\frac{\partial}{\partial t}\left(\rho \varphi\right) + \frac{\partial}{\partial x_j}\left(\rho U_j \varphi\right) = \frac{\partial}{\partial x_j}\left(\Gamma_{\text{eff}}\left(\frac{\partial \varphi}{\partial x_j}\right)\right) + S_\varphi \tag{3.5}$$

ANSYS CFX [18] software is utilized for the computations. A computer cluster and a work station are used for the simulations (the properties of which are given in Table 3.3).

3.3 RESULTS OF THE CASE STUDIES

3.3.1 CFD-AIDED ANALYSIS OF GASKETED PLATE HEAT EXCHANGERS AND EXPERIMENTAL VALIDATION

The main objective of CFD analysis is to shorten the design process time for gasketed plate heat exchangers, which is much less time-consuming than experimental procedures, and to determine the performance characteristics of newly designed plates by computer-aided methods. There are several approximations in the computational analysis; therefore, it is always necessary to verify the CFD model at hand. Figure 3.1 shows an appropriate verification process flow diagram for the accuracy estimations of the CFD analysis.

In Figure 3.8, the "Has the desired accuracy been reached?" step is used as a decision-making mechanism for validation. If the desired accuracy is not reached,

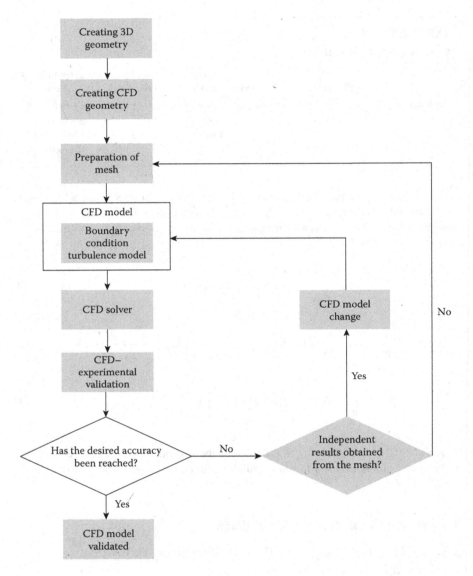

FIGURE 3.8 Flow chart for the CFD analysis.

first, mesh independency is obtained, then the turbulence model is revised for better results. However, with the help of the information gathered from the literature survey, the turbulence model decision is made before the mesh study.

3.3.1.1 Prediction of Hydraulic and Thermal Performance

In the experimental study, thermocouples are used to measure stable temperature changes for determining thermal characteristics of the plate. In this process, static temperature of the fluid is measured from a certain distance before it enters the heat exchanger. Therefore, in the CFD analysis, the same distances from the inlet and outlet surfaces are used and mean temperature differences are utilized in order to obtain thermal characteristics.

For the hydraulic performance evaluation, pressure differences are measured by a similar technique. Differential pressure gauges are positioned on the same plane with the thermocouples, and pressure drop values for static pressures are recorded every 20 s. Likewise, in the CFD analysis, static pressure differences are used to assess hydraulic performance, and the results are validated with the experimental findings.

3.3.1.2 Turbulence Modeling

As shown in the flow chart for CFD validation, a turbulence model study is performed. The results of the turbulence model study are shown in Table 3.4 and Figures 3.9 and 3.10. Turbulence model evaluation is carried out for a mass flow rate of 0.03 kg/s and a Reynolds number of 708 for 20.6°C temperature difference and 2.8 kPa pressure drop, which replicates a case for the experiments.

The contours for temperature and pressure obtained from the numerical results in Table 3.1 are presented in Figures 3.9 and 3.10. The limits of temperature and pressure changes for both cold and hot sides of the plates are the same for all turbulence models; these are 20°C–33°C for the cold side and 52°C–65°C for the hot side and 0–0.45 kPa for both hot and cold sides.

The numerical data from Table 3.4 and the temperature contours presented in Figure 3.9 indicate two main conclusions about turbulence characteristics from the k-ω and k-ε turbulence model families. Temperature difference estimations of standard k-ε, RNG k-ε, and EARSM k-ε models are below the expected values, and for standard k-ω and SST k-ω turbulence models, temperature differences are higher and more realistic. Temperature distribution gradually changes from the inlet to the outlet of both the hot and cold sides of the plate.

Pressure distributions obtained from different turbulence models are similar to the temperature distributions. As expected, the k-ω turbulence model family better demonstrates pressure drops resulting from eddies on the corrugations of the plate and has a better solving mechanism for near-wall functions and boundary layers.

TABLE 3.4

Results of Turbulence Model Studies for Re = 708

Turbulence Model	ΔT (°C)	ΔP (kPa)
k-ω	12.73	0.39
SST	12.73	0.39
k-ε	8.0	0.32
RNG k-ε	8.1	0.32
EARSM k-ε	8.8	0.34

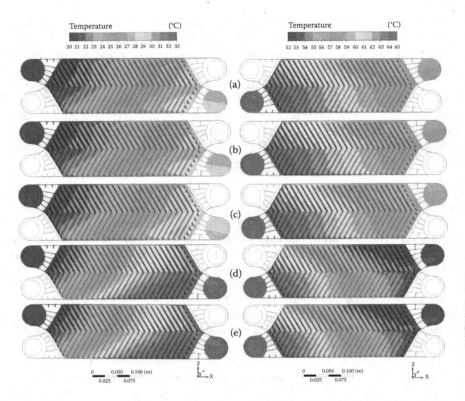

FIGURE 3.9 Temperature distribution of hot side (left) and cold side (right) plates: (a) k-ε, (b) RNG k-ε, (c) EARSM k-ε, (d) k-ω, and (e) SST k-ω turbulence models.

The SST k-ω turbulence model displays the smoothest gradual transition for both pressure and temperature distributions.

3.3.1.3 Grid Refinement Study

The number of elements is increased to a maximum based on computational facilities. Prismatic layers are added to the wall surfaces for better accuracy near the boundaries. Consequently, average y+ values obtained from these layers are investigated to refine the mesh. For this purpose, without adding prismatic layers, the number of elements are doubled iteratively, as shown in Table 3.5. In this table, the symbol M indicates million. For the first mesh, 2.5 million elements are used and then are doubled to 5 million, 10 million, 20 million, and 40 million elements. With the results gathered from the mesh study, the mesh containing 20 million elements is selected for the final computations instead of the mesh with 40 million elements in order to reduce the calculation cost. As in the turbulence model studies, Reynolds number, mass flow rate, temperature, and pressure drop values are kept the same. By increasing the number of elements in the mesh, the error in temperature drop decreases; however, pressure drop almost remains the same. Relative error

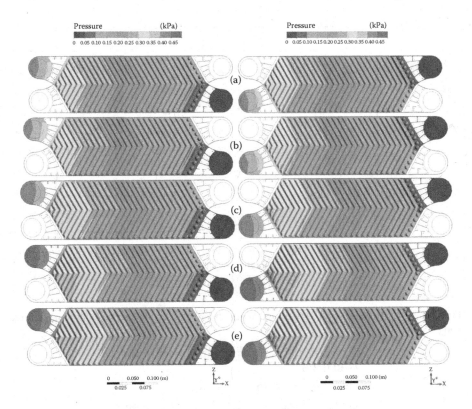

FIGURE 3.10 Pressure distribution of hot side (left) and cold side (right) plates: (a) k-ε, (b) RNG k-ε, (c) EARSM k-ε, (d) k-ω, and (e) SST k-ω turbulence models.

TABLE 3.5
The Effect of Number of Elements for Re = 708

Number of Elements	ΔT (°C)	ΔP (kPa)
2.5 M	12.7	0.39
5 M	14.9	0.40
10 M	17.0	0.40
20 M	19.6	0.43
40 M	19.9	0.44

definitions are given in Equations 3.6 and 3.7, and errors with respect to number of elements in the mesh are shown in Figure 3.11.

$$e_{\Delta T} = \frac{\left|\Delta T_{\text{CFD}} - \Delta T_{\text{Experiment}}\right|}{\Delta T_{\text{Experiment}}} * 100 \tag{3.6}$$

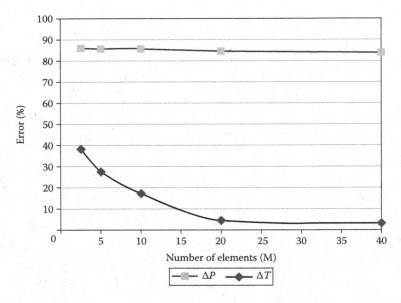

FIGURE 3.11   Relative error with respect to number of elements in the mesh for temperature and pressure drop.

TABLE 3.6
The Effect of y⁺ for Re = 708

Number of Elements	y+	ΔT (°C)	ΔP (kPa)
10 M	5	17.0	0.40
	1	17.8	0.46
	0.5	27.2	0.6
20 M	4.2	19.6	0.43
	0.01	22.1	1.17

$$e_{\Delta P} = \frac{\left|\Delta P_{\text{CFD}} - \Delta P_{\text{Experiment}}\right|}{\Delta P_{\text{Experiment}}} * 100 \tag{3.7}$$

The lack of difference in pressure drop errors with increasing number of elements requires reconsideration of y⁺ values. Therefore, prismatic layers are used in 10 million and 20 million elements to achieve consistent results with experiments. The y⁺ study results are given in Table 3.6.

It can be seen that with decreasing y⁺ values, pressure drop also increases. However, for thermal performance, with decreasing y⁺ values, transition to nonstructural elements exists, and there is a larger difference in temperature in the grid with 10 million elements, which is a result of the accumulation of numerical errors. For this reason, with the decreasing size of the elements near the boundary layer, a reduction in the size of the elements in the entire volume is needed resulting in increasing the number of elements; therefore, the final CFD study is conducted with the mesh that has 20 million elements. It is also observed that the turbulence models in the k-ω family give better results for near-wall solutions than those in the k-ε family.

3.3.1.4 Validation of Results Using Experimental Data

Three plates containing two flow channels are analyzed through the Nusselt number and friction factor correlations that are obtained from experimental results. Figure 3.12 shows the relative errors in temperature and pressure drop data of CFD results with respect to experimental results. The temperature drop values are compatible with experimental results, whereas the pressure drop values deviate from the experimental findings.

The average error value of the pressure drop values shown in Figure 3.12 is 60.8%, which is approximately 2.5 times that of the experimental results. Ismail and Velraj [9] suggested that a correction coefficient can be used in the design processes of CFD simulations of heat exchangers. Therefore, pressure drop values are multiplied by 2.5 and the comparison between the experimental results and CFD simulations is repeated. The errors in temperature and corrected pressure drop data of CFD results with respect to experimental results for a Reynolds number range of 708–5420 and mass flow rates varying from 0.03 kg/s to 0.24 kg/s are shown in Figure 3.13.

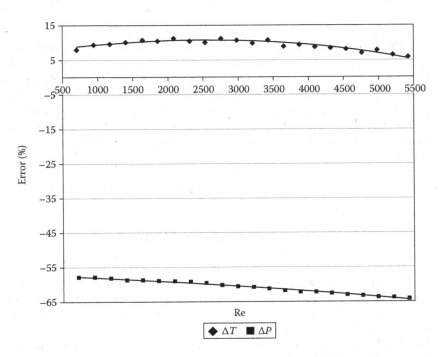

FIGURE 3.12 Errors in temperature and pressure drop data of CFD results compared to experimental results.

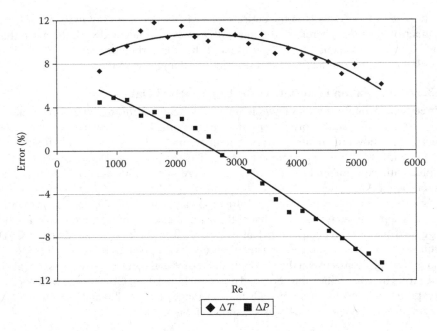

FIGURE 3.13 Errors in temperature and corrected pressure drop data of CFD results with respect to Reynolds number.

The CFD results indicate that maximum and minimum errors in temperature difference are 6% and 11.1% (9.3% average). When the Reynolds number exceeds 2755, the error in temperature drop decreases. Pressure drop values are higher than experimental results between the Reynolds numbers 708 and 2755, but after 2755, they show a decreasing trend. The maximum and minimum relative errors of pressure drop data are 0.56% and 10.4%, with an average value of 4.7%. The second-degree correlations that give errors in temperature and pressure differences are given in Equations 3.8 and 3.9.

$$\Delta T_{err}(\text{Re}) = -6*10^{-7}*\text{Re}^2 + 0.003*\text{Re} + 7.14 \tag{3.8}$$

$$\Delta P_{err}(\text{Re}) = -2*10^{-7}*\text{Re}^2 - 0.002*\text{Re} + 7.18 \tag{3.9}$$

Three different CFD simulations for validating the results are shown in Figures 3.14 and 3.15. Figure 3.14 shows temperature distribution, and Figure 3.15 shows pressure distribution. In these figures, the main variable is mass flow rate, which is 0.04 kg/s, 0.14 kg/s, and 0.24 kg/s.

Fluid temperature in distribution channels stays the same (in contrast to the flow channels) when heat transfer starts between the two sides of the plates. Also, most of the heat transfer takes place in the middle of the channels rather than in the parts closer to the

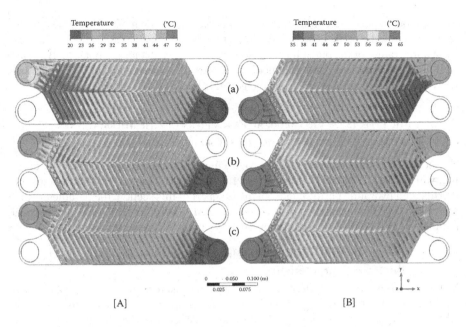

FIGURE 3.14 Temperature distribution of heat transfer plate, cold side [A], hot side [B]: (a) Re = 940, (b) Re = 3200, and (c) Re = 5420.

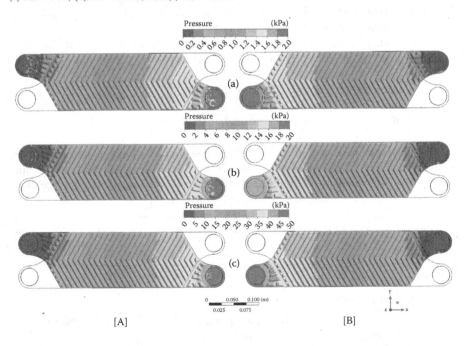

FIGURE 3.15 Pressure distribution of heat transfer plate, cold side [A], hot side [B]: (a) Re = 940, (b) Re = 3200, and (c) Re = 5420.

inlet and outlet ports. According to the simulation results, the most efficient heat transfer is in the middle of the plate, so even higher temperature values are achieved than the local temperature of the outlet for the cold side and vice versa for the hot side. By using well-arranged turbulence and grid models, the local temperature and pressure distributions are predicted without changing the general characteristics of plates.

3.3.2 The Effects of Geometrical Parameters on the Thermal and Hydraulic Characteristics of Gasketed Plate Heat Exchangers

The next step in the CFD-aided plate heat exchanger design process is to use the existing, validated CFD model to observe the effects of several geometrical parameters on thermal and hydraulic performance. In the first part of this section, the parameters that affect the flow characteristics will be determined. The effects of these geometrical parameters will be investigated computationally in the following sections.

3.3.2.1 Determination of the Geometrical Parameters that Affect the Flow Characteristics

Wave profile, wave amplitude, channel height, the distance between the plates, and distribution channels are important geometrical properties that determine the performance of plate heat exchangers. Wave amplitude, channel height, and distribution channel properties are selected for the design of new heat transfer plates. In the following sections, the effects of these parameters on thermal and hydraulic performance will be investigated in detail for a Reynolds number range of 708–5420. Table 3.7 shows these parameters in a representative channel.

The following values are used for wave amplitude and channel height in the simulations for plate heat exchangers with and without distribution channels.

- Wave amplitude: 2.0 mm, 2.25 mm, 2.75 mm
- Channel height: 2.5 mm, 3.0 mm
- Distribution channel: without channel, with channel

The plate geometries with and without distribution channels are shown in Figure 3.16. Table 3.7 shows the names and properties of the plates used in the simulations. Plate 7 is the plate without distribution channels, whereas Plate 8 is the plate with the distribution channels, as shown in Figure 3.16.

The temperature difference and pressure drop values obtained with each plate are compared to that of Plate 1, which is the standard commercial plate, whereas the

TABLE 3.7
Plate Properties

	Plate 1	Plate 2	Plate 3	Plate 4	Plate 5	Plate 6
Wave amplitude (mm)	2.5	2.25	2.00	2.75	2.5	2.5
Channel height (mm)	2.76	2.76	2.76	2.76	2.50	3.00

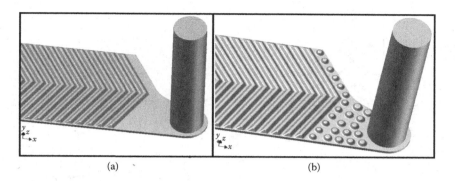

FIGURE 3.16 Plate geometries (a) without distribution channel and (b) with distribution channel.

other seven are the new plate designs. Equations 3.10 and 3.11 show the definitions of temperature and pressure differences compared to Plate 1.

$$\%\Delta T = \frac{\Delta T_{\text{Plate}-i} - \Delta T_{\text{Plate}-1}}{\Delta T_{\text{Plate}-1}} \times 100 \tag{3.10}$$

$$\%\Delta P = \frac{\Delta P_{\text{Plate}-i} - \Delta P_{\text{Plate}-1}}{\Delta P_{\text{Plate}-1}} \times 100 \tag{3.11}$$

3.3.2.2   The Effects of Wave Amplitude on the Flow Characteristics

The waviness of the plates is the major contribution to turbulence for the plates. The effect of the amplitude of the waviness of the surface is investigated numerically by considering several wave amplitudes for the same plate geometry. The wave amplitude of Plate 1 is increased from 2 mm to 2.75 mm gradually to observe the effects on the plate's thermal and hydraulic performance. Four different wave amplitudes are tested. The properties of the plates tested are given in Table 3.7. The changes in pressure drop and temperature difference values for four plates are shown in Figures 3.17 and 3.18. The differences in the results when compared to Plate 1 (existing commercial plate) are shown in Figure 3.19.

When the presented figures are examined, it is observed that the thermal performance drops with decreasing wave amplitude. However, the pressure drop values also decrease with decreasing wave amplitude. Wave shape helps vortices to develop, which affects the general behavior of the flow. When the wave amplitudes are lower, vortex development is prevented. Vortex development affects the thermal performance positively by increasing the vortices and therefore the turbulence; however, this also increases the pressure drop, which affects the hydraulic performance negatively. The enlargement coefficient is the ratio of the whole heat transfer surface to the smooth plate surface. The new enlargement coefficients for each of the new plates are shown in Table 3.8. Waviness helps turbulence development as well increasing the enlargement factor.

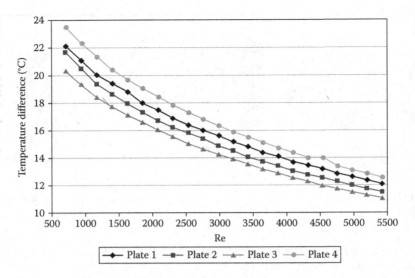

FIGURE 3.17 Variation of temperature difference with Reynolds number for plates with different wave amplitudes (Plate 1: 2.5 mm, Plate 2: 2.25 mm, Plate 3: 2.00 mm, and Plate 4: 2.75 mm).

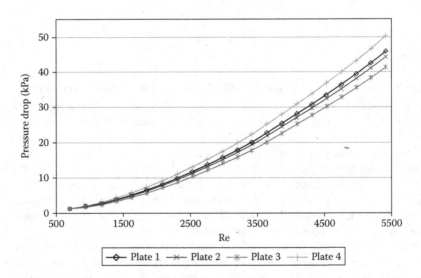

FIGURE 3.18 Variation of pressure drop with Reynolds number for plates with different wave amplitudes (Plate 1: 2.5 mm, Plate 2: 2.25 mm, Plate 3: 2.00 mm, and Plate 4: 2.75 mm).

3.3.2.3 The Effects of Channel Height on the Flow Characteristics

Channel height is an important factor that affects the thermal and hydraulic performance of plate heat exchangers. As shown in Table 3.7, Plates 5 and 6 have the same wave amplitude (2.5 mm) as Plate 1; however, the channel height for Plate 5 is 2.5 mm, and the channel height for Plate 6 is 3 mm. The results for temperature

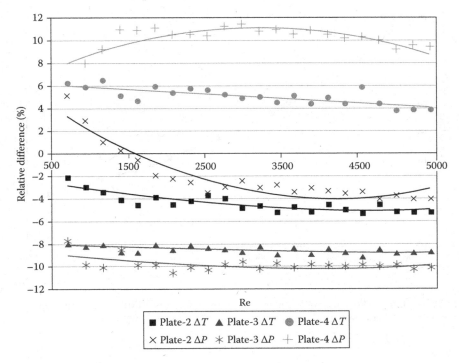

FIGURE 3.19   Temperature difference and pressure drop relative to Plate 1.

TABLE 3.8
Enlargement Factors for the Plates with Different Wave Amplitudes

	a (Wave Amplitude)(mm)	φ (Enlargement Factor)
Plate 1	2.50	1.153
Plate 2	2.25	1.134
Plate 3	2.00	1.117
Plate 4	2.75	1.170

difference and pressure drop are shown in Figures 3.20 and 3.21. When the channel height decreases, the flow is forced to flow in a narrow area; therefore, it is affected more from the waviness, which makes the vortices affect the flow more.

Both the temperature difference and the pressure drop increase when the channel height decreases, as shown in Figures 3.20 and 3.21, which leads to an increase in turbulence. The trend is the same with Plate 1. When Plate 5 is compared to Plate 1, its thermal performance is 13.4% higher; however, the pressure drop values are almost 50% higher. When the channel height decreases, the pressure drop values increase drastically.

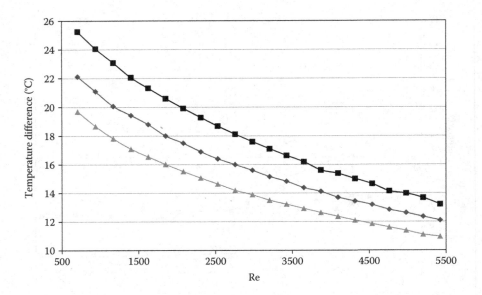

FIGURE 3.20 The effect of channel height on temperature difference with respect to Reynolds number (Plate 1: middle, Plate 5: top, and Plate 6: bottom).

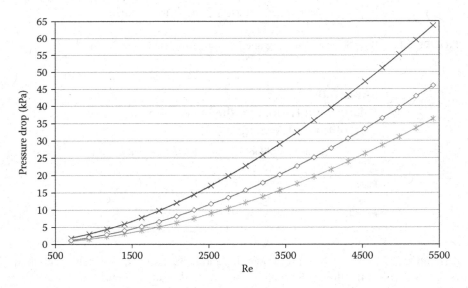

FIGURE 3.21 The effect of channel height on pressure drop with respect to Reynolds number (Plate 1: middle, Plate 5: top, and Plate 6: bottom).

Figure 3.22 shows the temperature difference for hot and cold channels for a Reynolds number of 1850 for three plates. The improvement in thermal performance with increasing channel height is also seen visually from the temperature distribution inside the channels.

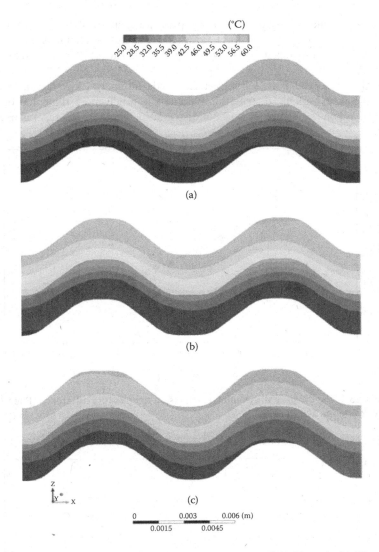

FIGURE 3.22 Temperature distribution inside the channel (a) Plate 6, (b) Plate 1, and (c) Plate 5.

3.3.2.4   The Effects of Distribution Channels on the Flow Characteristics

It is important for gasketed plate heat exchangers that the flow is distributed evenly in the channel so that all of the surface of the plates contribute to heat transfer. For this purpose, distribution channels that lead the flow from the inlet ports may be utilized. Investigating the effect of distribution channels on the flow structure usually has been ignored in the literature. As explained previously, among the CFD cases investigated, Plate 7 does not have distribution channels, whereas Plate 8 does.

The results for temperature difference and pressure drop are shown in Figures 3.23 and 3.24. As is seen in both of the figures, the distribution channels do not affect the average characteristics of the flow very much in terms of thermal and hydraulic performance.

It is also observed that, for higher Reynolds numbers (flow rates), the pressure drop because of the distribution channels is much more than that of lower flow rates. When the geometry without distribution channels (Plate 7) is used, the

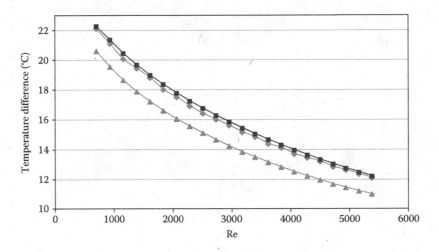

FIGURE 3.23   The effect of distribution channels on temperature difference (Plate 1: middle, Plate 7: bottom, and Plate 8: top).

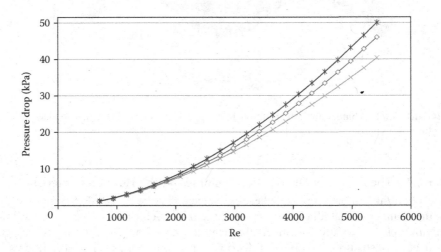

FIGURE 3.24   The effect of distribution channels on pressure drop (Plate 1: middle, Plate 7: bottom, and Plate 8: top).

average thermal performance drops by 8.5%, whereas the hydraulic performance drops by 6.8%.

3.4 DISCUSSION AND CONCLUSIONS

CFD analyses of a gasketed plate heat exchanger are performed and the results are validated using experimental findings. The SST k-ω turbulence model is used in the computations with a mesh size of 20 million elements—after careful analysis of the turbulence models and mesh sizes that can be used in the study. The experimentally verified CFD methodology is then further used to study the effects of plate geometrical parameters on the thermal and hydraulic numerical performances of the heat exchangers. The temperature differences for the gasketed plate heat exchangers obtained from the numerical analyses are in good agreement with the experimental findings. However, when the findings for pressure differences are compared with the experimental results, a discrepancy is observed. In order to compensate for this discrepancy, the pressure differences obtained from the numerical analyses are multiplied by a correction factor of 2.5.

Wave amplitude, channel height, and the distribution channels are important for plate design. These parameters are changed in a controlled manner to observe the effects of each factor on the performance of the plates. Wave profile increases the effective heat transfer area and turbulence; therefore, it increases the thermal performance. In order to investigate the effects of wave amplitude, two low amplitude plates and one high amplitude plate are designed and tested numerically. It is found that the thermal performance of Plates 2 and 3 are worse than that of Plate 1, which has a higher amplitude. On the other hand, the hydraulic performance based on pressure drop is better for low amplitude plates. Plate 4, which has the highest amplitude, is 5% better in thermal performance than the others; however, its pressure drop is 11% higher. This shows that, when the complexity of the plates is increased with the help of the patterns on it by increasing the total heat transfer area, the thermal performance improves, which also increases the pressure drop values and results in poor hydraulic performance. For applications where the thermal loads are more important, more complicated plate patterns can be selected instead of increasing the number of plates or using larger plates.

When the studies in the related literature are examined, channel height is found to increase thermal performance; however, it comes with poorer hydraulic performance based on the pressure drop values. By using CFD analyses of the current study, we found that, when we increase the channel height, the thermal performance drops 11% and the pressure drop values are 23% lower. If the decrease in thermal performance is tolerable, it is shown that the pressure drop values decrease drastically. It is also found that when the channel height is decreased, the thermal performance increases by 13%, whereas the pressure drop values increase by 48%.

Additionally important are the distribution channels; when the distribution channels are totally eliminated, the thermal performance decreases by about 9%. It is also found that the decrease in the thermal performance stays around the same percentages for the whole Reynolds number ranges. For the geometry with no channels, the flow is not distributed evenly, especially around the corners, and this is demonstrated by a decrease in thermal performance.

As a result, it is determined that there needs to be a compromise between thermal and hydraulic performance. Higher pressure drop values must be compensated when there is a need for thermal performance. For applications where a loss in thermal performance is tolerable, it is possible to decrease the pressure drop; therefore, energy is needed to pump into the system. When the results of the computational study are examined, it is observed that channel height is the most important parameter that affects both the thermal and hydraulic performance of plate heat exchangers. This study also shows that CFD can be a powerful and efficient tool for plate heat exchanger design, once the model is validated with the help of experiments that cannot totally be avoided.

NOMENCLATURE

ΔP	Pressure difference
ΔT	Temperature difference
Δt	Time step
A	Wave amplitude
B	Channel height
E	Relative error
F	Friction factor
J	Colburn factor
K	Turbulent kinetic energy
M	Dynamic viscosity
Nu	Nusselt number
Pr	Prandtl number
Re	Reynolds number
T	Time
V	Volume
ε	Turbulent dissipation
P	Density
Φ	Enlargement factor
Ω	Turbulent frequency

ACKNOWLEDGMENTS

This study is supported by Tubitak (the Turkish Scientific and Research Council) under grant number 112M173. The experimental facility used for the validation of the numerical results was funded by a grant from Turkish Ministry of Science, Technology, and Industry.

REFERENCES

1. S. Kakac, H. Liu, and A. Pramuanjaroenkij, *Heat Exchangers—Selection, Rating and Thermal Design*, 3rd Edition, CRC Press, Boca Raton, FL, 2002.

2. R.K. Shah, Classification of Heat Exchangers. In *Heat Exchangers: Thermal-Hydraulic Fundamentals and Design*, S. Kakaç, A. E. Bergles, and F. Mayinger (Eds.), Hemisphere, New York, pp. 9–46, 1981.

3. L. Wang, B. Sunden, and R.M. Manglik, *Plate Heat Exchangers Design, Applications and Performance*, WIT Press, Boston, MA, 2007.

4. S. Kakac, A. Pramuanjaroenkij, and H. Liu, *Heat Exchangers: Selection, Rating, and Thermal Design*, CRC Press, Boca Raton, FL, 2012.

5. S. Kakac, S. Aradag, and C. Gulenoglu, Experimental Investigation of Thermal and Hydraulic Performances of Gasketed Plate Heat Exchangers, *Engineer and Machinery*, Vol. 54, pp. 44–68, 2013. (In Turkish)

6. C. Gulenoglu, S. Aradag, N. Sezer-Uzol, and S. Kakac, Experimental Comparison of Performances of Three Different Plates for Gasketed Plate Heat Exchangers, *International Journal of Thermal Sciences*, Vol. 75, pp. 249–256, 2014.

7. F. Akturk, N. Sezer-Uzol, S. Aradag, and S. Kakac, Experimental Investigation and Performance Analysis of Gasketed-Plate Heat Exchangers, *Journal of Thermal Science and Technology*, Vol. 35, no. 1, pp. 43–52, 2015.

8. H. Blomerius and N.K. Mitra, Numerical Investigation of Convective Heat Transfer and Pressure Drop in Wavy Ducts, *Numerical Heat Transfer, Part A*, pp. 37–54, 2000.

9. L.S. Ismail and R. Velraj, Studies on Fanning Friction (f) and Colburn (j) Factor of Offset and Wavy Fins Compact Plate Fin Heat Exchanger -A CFD Approach, *Numerical Heat Transfer, Part A*, pp. 987–1005, 2009.

10. O. Pelletier, F. Stromer, and A. Carlson, CFD Simulation of Heat Transfer in Compact Brazed Plate Heat Exchangers, *ASHRAE Transactions*, Vol. 111, no. 1, pp. 846–854, 2005.

11. R.Y. Miura, F.C.C. Galeazzo, C.C. Tadini, and J.A.W. Gut, The Effect of Flow Arrangement on the Pressure Drop of Plate Heat Exchangers, *Chemical Engineering Science*, Vol. 63, pp. 5386–5393, 2008.

12. S. Jain, A. Joshi, and P.K. Bansal, A New Approach to Numerical Simulation of Small Sized Plate Heat Exchangers with Chevron Plates, *Journal of Heat Transfer*, Vol. 129, pp. 291–297, 2007.

13. S. O'Halloran and A. Jokar, CFD Simulation of Single-Phase Flow in Plate Heat Exchangers, *ASHRAE Transactions*, LV-11-C018, pp. 147–156, 2011.

14. A.G. Kanaris, A.A. Mouza, and S.V. Paras, Flow and heat tranfer in narrow channels with corrugated walls: a CFD code application, *Chemical Engineering Research and Design*, Vol. 83, no. 5, pp. 460–468, 2005.

15. W. Han, K. Saleh, V. Aute, G. Ding, Y. Hwan, and R. Radermacher, Numerical Simulation and Optimization of Single-Phase Turbulent Flow in Chevron-Type Plate Heat Exchanger with Sinusoidal Corrugations, *HVAC&RESEARCH*, Vol. 17, no. 2, pp. 186–197, 2011.

16. V. Patil, H. Manjunath, and B. Kusammanavar, Validation of Plate Heat Exchanger Design Using CFD, *International Journal of Mechanical Engineering and Robotics Research*, Vol. 2, no. 4, pp. 222–230, 2013.

17. L. Zhang and D. Che, Turbulence Models for Fluid Flow and Heat Transfer between Cross Corrugated Plates, *Numerical Heat Transfer, Part A*, Vol. 60, pp. 410–440, 2011.

18. CFX ANSYS Inc., CFX 14.0 Theory Guide, 2011. www.ansys.com

4 Numerical Methods for Micro Heat Exchangers

Bengt Sundén, Zan Wu, and Mohammad Faghri

CONTENTS

ABSTRACT: This chapter presents a state-of-the-art overview of numerical methods for single-phase flow and two-phase flow in microchannel heat exchangers, as well as the relevant single-phase and two-phase computational fluid dynamics (CFD) applications. Governing equations are given for both single-phase flow and two-phase flow with and without phase change. For single-phase flow, scaling effects, such as conjugate heat transfer and viscous heat dissipation, are considered for their significance on fluid flow and heat transfer in micro heat exchangers. For two-phase flow, the challenges of numerical modeling and the relative magnitudes of the dominant forces are discussed. Characteristics of the multiphase flow modeling approaches (i.e., the Eulerian–Eulerian method, the Eulerian–Lagrangian method, and direct numerical simulation [DNS]) are compared. The advantages and disadvantages of several continuum DNS methods for interface evolution (e.g., volume of fluid [VOF], level set, phase field, front-tracking, and moving mesh methods) and the mesoscopic lattice Boltzmann method (LBM) are discussed.

Methods to address the mass nonconservation in the level set method are briefly provided. Because microchannels are the basic elements of micro heat exchangers, recent CFD applications of two-phase flow in microchannels, mostly limited to the scale of a few bubbles or droplets, are briefly summarized. Future research needs for numerical modeling of micro heat exchangers are suggested.

4.1 INTRODUCTION

Micro heat exchangers (μHEXs) are heat exchangers in which at least one fluid flows in lateral confinements such as microchannels or tubes with typical dimensions less than 1 mm. They have been fabricated with high heat transfer coefficients approximately one order of magnitude higher than the typical values for conventional heat exchangers. Therefore, as a miniaturized process device, μHEXs have attracted widespread applications due to their high thermal performance, compactness, and their small size and weight [1]. Since the pioneering work of Tuckerman and Pease [2] in 1981, which exhibited for the first time a high heat-flux dissipation of up to 790 W/cm^2 by using microchannels, a great deal of work has been devoted to investigating single-phase and two-phase fluid flow and heat transfer characteristics in μHEXs [3].

Micro heat exchangers are used in diverse energy and process applications such as electronics cooling, automotive and aerospace industries, chemical process intensification, refrigeration and cryogenic systems, and in fuel cells [3–9]. They have many advantages over conventional heat exchangers. First, in the scaling down from macro to microscale, the volume decreases with the third power of the characteristic linear dimensions, while surface area only decreases with the second power. Therefore, μHEXs have relatively larger surface area to volume ratios that enable higher heat transfer rates than conventional heat exchangers. Second, high fluid acceleration and close proximity of the bulk fluid to the wall surface in μHEXs give high heat transfer coefficient values. For single-phase fully developed internal laminar flows, a constant Nusselt number (Nu = hd_h/k) implies that the heat transfer coefficient increases as the hydraulic diameter decreases. In addition, compactness and high heat-flux dissipation are required as the scale of the devices becomes smaller while the power density becomes larger. Micro heat exchangers are expected to have high heat-flux dissipation for miniaturized devices to guarantee reliability and safety during operation.

Although the development of μHEXs mainly has been driven by demands related to industrial process intensification (with reduced inventories and reduced footprints) and a sustainable environment, progress in the area of material science has contributed to the manufacture of μHEXs. Both traditional and modern micromachining techniques were applied to fabricate μHEXs [10]. Traditional techniques such as computer numerical control (CNC) mill, micro sawing, dicing, and micro deformation are possible choices. Modern fabrication techniques include laser micromachining; deep reactive ion etching; Lithography, Electroplating, and Molding (Lithographie, Galvanoformung, Abformung; LIGA); wafer bonding; diffusion bonding; electrodischarge machining; and three-dimensional (3D) printing. Advantages and disadvantages of several micromachining methods were given in Sundén and Wu [10]. Figure 4.1 shows several examples of fabricated μHEXs.

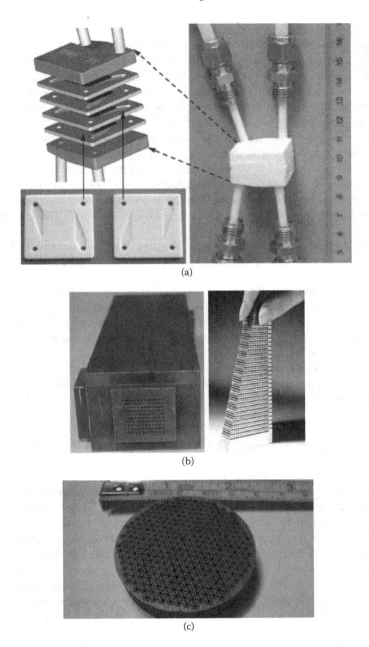

(a)

(b)

(c)

FIGURE 4.1 Examples of μHEXs: (a) ceramic micro heat exchanger (From Alm, B., et al., *Chem. Eng. J.*, 135, S179–S184, 2008. With permission.), (b) diffusion bonded Alloy 617 heat exchanger (From Mylavarapu, S.K., et al., *Nucl. Eng. Des.*, 249, 49–56, 2012. With permission.), and (c) 3D-printed heat exchanger by Sustainable Engine Systems Ltd. (From Reay, D., et al., *Process Intensification: Engineering for Efficiency, Sustainability and Flexibility*, Butterworth-Heinemann, Oxford, 2013. With permission.)

The transport phenomena occurring in μHEXs with small channels/confinements can be significantly different from that of conventional macroscale heat exchangers. In the present work, the flow regimes for channels/confinements for single-phase flows and two-phase flows are defined separately. For gas flows, the Knudsen number, Kn, which is the ratio of the mean free path to the characteristic length of the channel, can best describe the flow regimes [10,13]. For Kn < 0.01, the flow is considered to be continuum. For 0.01< Kn < 0.1, the flow is in the slip flow regime where the continuum equations are valid with the application of the slip boundary conditions. The transition regime is present for 0.1 < Kn < 3 where the flow is too rarefied for continuum analysis, and the Boltzmann equation or its coarse-grained version, direct simulation Monte Carlo (DSMC), is applicable. For Kn > 3, the flow regime is considered to be in the free molecular regime where collisions among molecules are neglected and the collisionless Boltzmann equation can be used.

For single-phase liquid flows, the mean free path is not well defined. However, an equivalent Kn number can be calculated based on an approximate mean free path to determine the flow regimes.

Based on the aforementioned criteria, the continuum analysis is valid when hydraulic diameters are greater than 100 μm for gas flows and greater than 10 μm for single-phase liquid flows. For most practical applications, we can assume the hydraulic diameters of μHEX channels/confinements are in the range of ~10 μm $\leq d_h \leq$ ~1.0 mm for liquid flows.

For two-phase flows, especially for flow boiling, Wu and Sundén [14] proposed an upper limit for microscale channels/confinements in μHEXs for both normal gravity and microgravity conditions:

$$BoRe_1^{0.5} = 200 \text{ and } Bo = 4 \tag{4.1}$$

where Bo and Re_1 are the Bond number and the liquid Reynolds number, respectively. The first part in Equation 4.1, $BoRe_1^{0.5} = 200$, validated in the literature [15–17], is intended to ensure bubble confinement at relatively large liquid Reynolds numbers. The second part in Equation 4.1, $Bo = 4$, proposed by Kew and Cornwell [18], can ensure bubble confinement at low liquid Reynolds numbers. Therefore, for two-phase flow, Equation 4.1 is set as the upper limit for μHEXs, and ~10 μm is set as the lower limit for microchannels. Classification of flow regimes in two-phase flow can also be based on the hydraulic diameters as well as fluid and flow conditions.

Computational fluid dynamics (CFD) is a numerical solution methodology of the governing equations for conservations of mass, momentum, energy, and other transport processes. Due to advances in CFD, numerical models have been proposed to address single-phase and two-phase flows, thus providing essential information on the local flow structure and performance improvement of heat exchangers. CFD can be applied to heat exchangers in a variety of ways [19]. In one method, the entire heat exchanger is modeled. This can be done by using large-scale or relatively coarse computational mesh or by applying the "distributed-resistance" concept. For the latter case, volume porosities, surface permeability, as well as flow and thermal resistances have to be introduced. Another method is to identify modules or group of modules that repeat themselves in a periodic or cyclic manner in the main flow direction. This will enable accurate calculations for the modules, but the entire heat exchanger—including manifolds and distribution areas—is not included.

Several commercial, general-purpose CFD codes have been used extensively, such as ANSYS FLUENT, ANSYS CFX, STAR-CD (STAR CCM+), TransAT, CFD-ACE+, and COMSOL Multiphysics. Recently, open source codes such as Open Foam also have been used. In addition, many in-house computational codes have been developed in various universities and research centers to numerically solve various problems.

The Knudsen number is a useful parameter for determining how to treat single-phase flows in μHEXs in terms of a continuum or noncontinuum approach. Because the *Kn* for many practical μHEXs is smaller than 0.01, the continuum approach is valid for most applications. The momentum and energy conservation equations can be solved with slip velocity and temperature boundary conditions to extend the existing continuum approaches to the finite Knudsen regime (e.g., the slip flow regime) with *Kn* up to 0.1 [20]. At a sufficiently large Kn (e.g., larger than about 0.1), noncontinuum kinetic models have to be implemented, such as the DSMC [21–23] for gases and the dissipative particle dynamics (DPD) [24,25] for liquids or the coarse-grained molecular dynamics [26]. For practical μHEXs with channel hydraulic diameters larger than 100 μm, continuum formulation and the mesoscopic lattice Boltzmann method (LBM) can be utilized.

Although numerical modeling for general single-phase flow and two-phase flow are the topic of several books [27,28] and review papers [19,29], only a few reviews [30] are specifically devoted to numerical modeling of single-phase and two-phase flows in μHEXs. The recent advances in single-phase and multiphase CFD techniques are able to provide new and detailed insights into the local hydrodynamics and thermal features in components of μHEXs. For two-phase flow, the accuracy of the phase interface evolution and modeling of interfacial effects is of primary importance for microscale-aimed computational methods because the interface topology plays a fundamental role in flows within micro devices.

This chapter aims to provide a comprehensive review of numerical modeling of μHEXs with emphasis on two-phase flows, as well as to present relevant CFD modeling for a variety of applications.

4.2 NUMERICAL MODELING OF SINGLE-PHASE FLOW IN μHEXs

4.2.1 GOVERNING EQUATIONS

The governing differential equations for conservations of mass, momentum, and energy can be cast into a general partial differential equation [31]:

$$\frac{\partial(\rho\varphi)}{\partial t} + \frac{\partial}{\partial x_i}(\rho\varphi u_i) = \frac{\partial}{\partial x_i}\left(\Gamma\frac{\partial\varphi}{\partial x_i}\right) + S \qquad (4.2)$$

where φ is an arbitrary dependent variable (e.g., unit, velocity components, or $c_p T$). Parameter Γ is the generalized diffusion coefficient, and S is the source term for φ. The general differential equation consists of four terms. From left to right in the Equation 4.2, they are known as the unsteady term, the convection term, the diffusion term, and the source term. The laminar convective flow and heat transfer can be simulated by Equation 4.2.

Turbulent flow and heat transfer normally require modeling approaches in addition to Equation 4.2. For turbulence modeling, the goal is to account for all of the relevant physics by using as simple a mathematical model as possible [19]. Commonly, a time-averaged operation called Reynolds decomposition is carried out. Every variable is then written as a sum of a time-averaged value and a superimposed fluctuating value. In the governing equations, additional unknowns appear, six for the momentum equations and three for the temperature field equation. The general variable is then written as:

$$\varphi = \overline{\varphi} + \varphi'$$ (4.3)

The additional terms in the differential equations have the forms

$$-\rho \overline{u_i' u_j'} \text{ and } \rho c_p \overline{u_i' T'}$$ (4.4)

and are called turbulent stresses and turbulent heat fluxes, respectively. The task of turbulence modeling is to provide procedures to predict the additional unknowns (i.e., the turbulent stresses and turbulent heat fluxes) with sufficient generality and accuracy. Because the flow in typical µHEXs is laminar, very few numerical studies included turbulence models such as the standard k-ε and low-Re k-ε models.

There are several methods available for numerical solution of the governing equations of fluid flow and heat transfer problems, such as the finite volume method [27,31], the finite difference method [32], the finite element method [33], and the boundary element method [34]. Interested readers may refer to the literature [31,32–34] for details of these numerical methods.

4.2.2 Scaling Effects for Single-Phase Flow in µHEXs

For single-phase flow in µHEXs with channel hydraulic diameters ranging from about 100 µm to 1 mm, the conventional theories and correlations are still able to predict the fluid flow and heat transfer characteristics in microchannels when the following effects are considered appropriately:

- Microchannel geometry
- Entry and exit losses
- Surface roughness
- Entrance effects
- Conjugate heat transfer
- Viscous dissipation
- Temperature-dependent properties

For hydraulic diameters less than about 100 μm for gases and 10 μm for liquids, additional scaling effects, such as electric double layer (EDL), rarefaction, compressibility, and roughness should be considered. These possible scaling effects, often negligible in conventional macro channels, may now have a significant influence for microchannels [35–39]. In this chapter, we will mainly focus on conjugate heat transfer and viscous dissipation. Interested readers may refer to a recent review by Dixit and Ghosh [1] on single-phase liquid flows in micro heat exchangers for the effect of EDL and Faghri et al. [13,20–25] for noncontinuum analysis using DSMC and DPD for the effects of rarefaction, compressibility, and roughness in microchannels.

Conjugate heat transfer refers to the ability to compute conduction of heat through solids, coupled with convective heat transfer in a fluid with coupled boundary conditions. For a μHEX with parallel microchannels, three-dimensional (3D) effects in the solid walls and in the fluid might significantly affect the heat transfer behavior. Maranzana et al. [40] introduced a dimensionless number (M) to evaluate whether conjugate heat transfer should be considered:

$$M = \frac{\text{axial heat conduction in the wall}}{\text{convective heat transfer in the fluid}} = \frac{k_s}{\rho_f c_{pf} u} \frac{A_s}{A_f L} = \frac{k_s}{k_f} \frac{A_s}{A_f} \frac{d_h}{L} \frac{1}{\text{Re Pr}} \quad (4.5)$$

where A_s and A_f indicate the cross-sectional areas of the solid and the fluid, respectively. If $M < 0.01$, the axial heat conduction in the solid walls can be neglected. Figure 4.2 shows the variation of M with RePr for a typical microchannel heat sink with a thermal conductivity ratio (k_s/k_f) of 350, a length to diameter ratio (L/d_h) of 50, and a cross-sectional area ratio (A_s/A_f) of 1. As shown in Figure 4.2, M becomes larger than 0.01 when RePr is less than around 700. Thus, conjugate heat transfer needs to be considered in the numerical modeling, especially at relatively low Reynolds numbers (~100 to ~700 depending on the Pr value) in laminar flow. Different from microchannel heat sinks, the wall thickness is comparably much smaller than the channel diameter for conventional channels. Therefore, the cross-sectional area ratio (A_s/A_f)

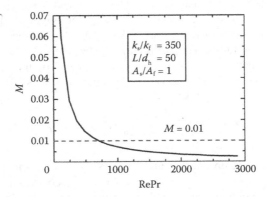

FIGURE 4.2 The variation of M with RePr for a typical microchannel heat sink. The gray dashed line indicates $M = 0.01$. When $M > 0.01$, the conjugate heat transfer should be considered in the numerical modeling.

is much less than 1 and the M value is less than 0.01 even at very low Re values for conventional channels.

In the conjugate heat transfer model involving simultaneous determination of the temperature field in both the solid and the liquid regions, an additional energy equation for the solid part should be solved together with the continuity equation, the momentum equations, and the energy equation for the fluid. The heat conduction in the solid section at steady state without internal heat generation can be stated as

$$\frac{\partial}{\partial x}\left(k_s \frac{\partial T_w}{\partial x} \right) + \frac{\partial}{\partial y}\left(k_s \frac{\partial T_w}{\partial y} \right) + \frac{\partial}{\partial z}\left(k_s \frac{\partial T_w}{\partial z} \right) = 0 \tag{4.6}$$

The continuity of the temperature and heat flux is used as conjugate boundary conditions to couple the energy equations for the fluid and solid phases; see for example, Li et al. [41], Sui et al. [42], and Tiselj et al. [43]. With conjugate boundary conditions at the liquid-solid interface and boundary conditions at the inlet and outlet, the fluid flow and heat transfer in the microchannel heat sink and μHEXs can be solved numerically.

To exemplify the viscous dissipation, the energy equation for an incompressible fluid is rewritten as follows

$$\rho_f c_p \frac{\partial T_f}{\partial t} + \rho_f c_p u_f \cdot \nabla T_f = \nabla \cdot \left(k_f \nabla T_f \right) + \psi \tag{4.7}$$

where ψ is the viscous dissipation that represents the irreversible conversion of mechanical energy to internal energy due to the deformation of fluid elements.

$$\psi = \frac{1}{2} \mu_f \nabla u_f \cdot \nabla u_f \tag{4.8}$$

The viscous dissipation is usually neglected in conventional channels. However, the internal heat generation due to viscous dissipation might produce a significant temperature rise in microchannels, even at low Reynolds numbers for high-viscous and low heat capacity fluids. The Brinkman number, Br, defined to be the ratio of heat produced by viscous dissipation to heat transported by molecular conduction, can be used to evaluate whether the viscous dissipation effects are important.

$$\text{Br} = \frac{\text{viscous dissipation}}{\text{thermal conduction}} = \frac{\mu_f u^2}{k_f |T_w - T_{\text{bulk}}|} \tag{4.9}$$

From Equation 4.9, at similar flow and operating conditions, the Brinkman number depends on fluid properties, especially the dynamic viscosity and thermal conductivity. The relationship between the Brinkman number and the Reynolds number is shown in Figure 4.3 for three common heat transfer fluids (i.e., water, ethylene glycol, and engine oil). The Br is larger than 1.0 for engine oil due to its very high viscosities. Thus, the viscous heat dissipation term cannot be neglected in the energy equation for

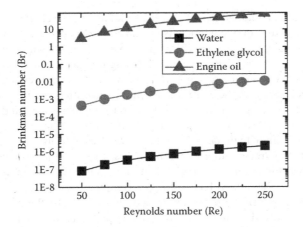

FIGURE 4.3 The relationship between the Brinkman number and the Reynolds number for three fluids (water, ethylene glycol, and engine oil) with a wall superheat of 30°C. The properties were evaluated at 50°C.

engine oil, even at very low Reynolds numbers down to 10. The Br for water is very low (~10^{-7}) due to its relatively low viscosity and high thermal conductivity. The viscous dissipation can be neglected during the numerical modeling of water in laminar flow in μHEXs. For ethylene glycol, the viscous dissipation might be neglected at low Reynolds numbers (~100); however, it should be included at relatively high Reynolds numbers (~1000).

Thermophysical properties—including density, heat capacity, thermal conductivity, and viscosity of fluids—might vary considerably with temperature. Relevant correlations for these temperature-dependent properties are listed in various sources. Piece-wise linear relationships can be considered for density, heat capacity, and thermal conductivity, while polynomial approaches can approximate the variation of viscosity with temperature. Temperature-dependent properties were considered in some numerical studies, such as Li et al. [41] and Xu et al. [44], and the results indicate that thermophysical properties of the liquid can significantly affect the flow and heat transfer in μHEXs.

4.2.3 CFD APPLICATIONS FOR SINGLE-PHASE FLOW

Entrance effects, conjugate heat transfer, viscous dissipation, thermophysical property variations, inlet and outlet plenums, and flow distribution need to be appropriately incorporated into the numerical models of μHEXs, depending on the specific μHEX applications. Contrary to conventional heat exchangers, asymmetrical heating and adiabatic conditions, which are very common in μHEXs, should also be considered in the boundary conditions. For simplicity, in most cases, only a short section or a single channel is simulated by using symmetrical and periodic assumptions, which might not be the case for real μHEXs; therefore, the results could not be extended to the whole μHEX. A complete model, including a silicon chip with 17 microchannels and inlet and outlet plenums/collectors, was simulated by Tiselj et al. [43]. However, the grid density of the complete model is far too coarse to accurately capture the temperature and velocity profiles inside the microchannels, as shown in Figure 4.4. Two main features are observed in Figure 4.4: (i) significant temperature gradients in the inlet and outlet plenums, which indicate that the temperatures measured by the thermocouples inserted in the centers of the inlet and outlet plenums are up to a few degrees lower than the actual inlet and outlet bulk temperatures, and (ii) the measured temperature difference between the inlet and outlet plenums is higher than that predicted by the complete model [43]. Fluid flow and heat transfer behaviors inside a counter-flow μHEX (rectangular microchannels with hydraulic diameter of 375 μm) were determined by a complete model using the COMSOL Multiphysics software in Dang et al. [45]. Pieper and Klein [46] proposed a numerical network flow model for bionic μHEXs generated by an optimized algorithm. The fluid dynamics in the 3D channels was solved by an LBM. The velocity and mass flow data from the LBM were then used as input for the network flow model. Because the LBM requires a high resolution, it is only possible to simulate times below a second. Using the network flow model, the temperature distribution at the network vertices was obtained. The designed bionic μHEXs can provide more uniform flow distribution and thus better heat transfer performance than conventional heat exchangers.

Complete models could present overall temperature and velocity distributions for μHEX. Possible optimization of the μHEX might be suggested based on the complete model. However, in order to design a μHEX or to further optimize the μHEX,

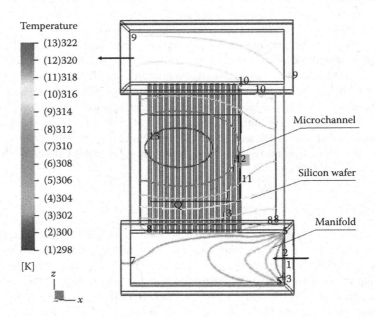

FIGURE 4.4 Temperature profile simulated by the complete model including a silicon chip with 17 microchannels and inlet and outlet manifolds. (Adapted from Tiselj, I., et al., *Int. J. Heat Mass Tran.*, 47, 2551–2565, 2004.)

models with smaller computational domain are required to obtain detailed fluid flow and heat transfer characteristics in the microchannels. Most relevant numerical studies are mainly specified for one microchannel or a part of the microchannel array. Miwa et al. [47] numerically investigated the heat transfer performance of two-stream parallel and counter-flow gas-to-gas µHEXs. Choquette et al. [48] presented an optimum design of microchannel heat sinks with optimal dimensions that minimize the heat exchanger's thermal resistance. Li et al. [41] conducted a numerical simulation of heat transfer in a microchannel heat sink with rectangular microchannels with a hydraulic diameter of 86 µm by using a simplified 3D conjugate heat transfer model (2D fluid flow and 3D heat transfer) with temperature-dependent thermophysical properties. The results showed that the bulk liquid temperature varies in a quasilinear form along the flow direction for high fluid flow rates but not for low flow rates.

At low flow rates, particular attention should be paid to heat losses associated with the longitudinal heat conduction from the substrate to the ambient. A log-mean temperature difference was suggested by Li et al. [41] to interpret heat transfer data consistently. Foli et al. [49] combined CFD analysis with an analytical method to optimize geometric parameters of µHEXs. There is a trade-off between minimum pressure drop and maximum heat transfer. The optimized µHEX yielded higher heat fluxes and heat transfer rates. Thermal performance and pressure drop of a double-layer microchannel heat sink for both parallel-flow layout and counter-flow layout were investigated numerically and optimized by Xie et al. [50]. One half of a double-layer microchannel was taken as the computational domain, as shown in Figure 4.5a. The fluid subdomain contains two half-channels, and the others belong to solid subdomains. Figure 4.5b shows the local static temperature distributions of the computational domain for parallel-flow and counter-flow, respectively, at an inlet velocity of 0.5 m/s. For parallel-flow, the temperature of the bottom surface increases along the flow direction with the maximum temperature located at the channel outlets. For counter-flow, the surface temperature increases first and then decreases along the flow direction. The maximum temperature is located close to the middle region. Xie et al. [50] suggested that the double-layer microchannel heat sink reduced the pressure drop and exhibited better thermal performance compared with a single-layer microchannel heat sink at the same pumping power. The parallel-flow layout is better for heat dissipation at low flow rates, while the counter-flow configuration is more efficient at high flow rates.

According to Wu and Sundén [3], the single-phase heat transfer in microchannel heat exchangers can be further enhanced by various passive enhancement techniques, such as flow interruptions, curved passages, and reentrant cavities. Xu et al. [44] proposed an interrupted microchannel heat sink by numerical simulations. Multichannel effect, entrance effect, physical property variations, and axial thermal conduction were considered in their 3D model. The computed velocity and temperature boundary layers are redeveloping in each separated zone in the interrupted heat sink and are responsible for the significant heat transfer enhancement. As seen in Figure 4.6, the Nusselt number in the interrupted microchannel heat sink is significantly high and has a decrease in each separated zone, but it is always larger than that for the conventional microchannel heat sink.

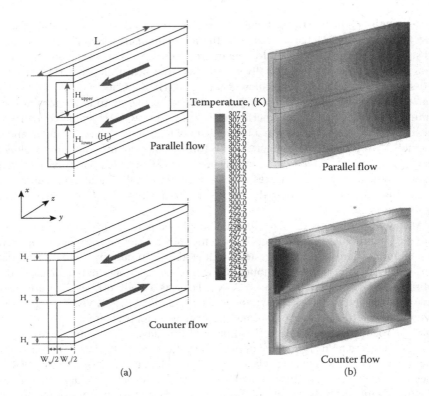

FIGURE 4.5 Numerical simulation of a double-layer microchannel heat sink: (a) the computational domain and (b) the static temperature contour at an inlet velocity of 0.5 m/s. (From Xie, G., et al., *ASME J. Therm. Sci. Eng. Appl.*, 5, 011004, 2013. With permission.)

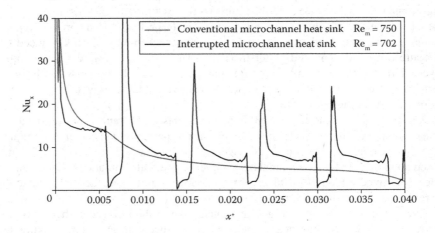

FIGURE 4.6 Cross-sectional average Nusselt number versus nondimensional flow length x^+ ($x/(d_h Re Pr)$) for conventional and interrupted microchannel heat sinks. (From Xu, J., et al., *Int. J. Heat Mass Tran.*, 51, 5906–5917, 2008. With permission.)

Sui et al. [42] simulated laminar liquid-water flow and heat transfer in wavy microchannels. Figure 4.6 shows the Nusselt distribution and the temperature distribution for two types of wavy microchannels. As shown in Figure 4.7a, the relative waviness increased along the flow direction for the left-side type, while for the right-side type, the relative waviness was designed to be higher at high heat-flux regions for hot spot mitigation purposes. Figure 4.7b presents the Nusselt number along the flow direction at Re = 400 under constant heat flux (H2), constant temperature (T), and conjugate heat transfer conditions for the two types of wavy microchannels.

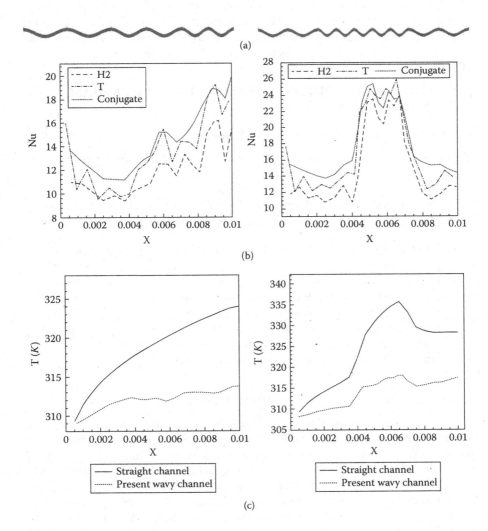

FIGURE 4.7 Numerical studies of two types of wavy microchannels: (a) illustration of the two types of wavy microchannels, (b) local Nusselt number along the flow direction at Re = 400, and (c) substrate temperature distribution along the flow direction at Re = 400. (From Sui, Y., et al., *Int. J. Heat Mass Tran.*, 53, 2760–2772, 2010. With permission.)

The Nusselt distributions are different at different wall boundary conditions. Generally, there is a significant heat transfer enhancement corresponding to the region with a higher relative waviness. The average span-wise temperature of the substrate along the flow direction is shown in Figure 4.7c. There is only a very moderate temperature rise for the wavy microchannels compared with the drastic temperature rise for the straight microchannel. Xie et al. [51] also numerically investigated the thermal performance of longitudinal and transversal-wavy microchannel heat sinks for electronic cooling. Fan-shaped reentrant cavities and internal ribs with different rib heights were investigated by Xia et al. [44] for Reynolds numbers ranging from 150 to 600. The combined effect of cavity and rib has better heat transfer performance than individual cavity, and the relative rib height presents a stronger effect than the arrangement or size of the reentrant cavities.

4.3   NUMERICAL MODELING OF GAS–LIQUID TWO-PHASE FLOW IN μHEXs

4.3.1   CHALLENGES OF MODELING GAS–LIQUID FLOW

There are several major challenges for modeling gas–liquid two-phase flow:

- High nonlinearity of governing equations.
- Vast ranges of time and length scales. Figure 4.8 describes the boiling and condensation processes according to their physical mechanisms, time scales, and length scales [52].
- Interface discontinuity: The phase interface separating the liquid from the gas is extremely thin.
- High density contrast: Density change across the phase interface is large.
- Phase interface exerts a localized surface tension force on the liquid.
- Phase transition (e.g., evaporation and condensation).
- Large interface deformations and topology changes (e.g., breakup and coalescence events associated with bubbles and droplets).
- Surface effects such as surface roughness and wettability.

4.3.2   DOMINANT FORCES IN GAS–LIQUID FLOW

Any fluid motion is controlled by forces. Basically, there are three categories of forces: volume forces (body forces), surface forces, and line forces. Five or six forces come into play for gas–liquid two-phase flow [53–55], such as pressure force, gravity (or buoyancy), inertia, viscous force, surface tension, and evaporation momentum force for boiling. Table 4.1 summarizes the dominant forces and their magnitudes in gas–liquid two-phase flow. Parameter L is the characteristic length. The evaporation momentum force, as described in Kandlikar [55], acts on the evaporating interface and plays a major role in its motion. The evaporation momentum is not present in nonboiling gas–liquid flows. As shown in Table 4.1, these forces have different dependences on the characteristic length. Therefore, the relative magnitudes of these forces may vary greatly with the characteristic length. That means, the relative magnitudes of these forces in μHEXs

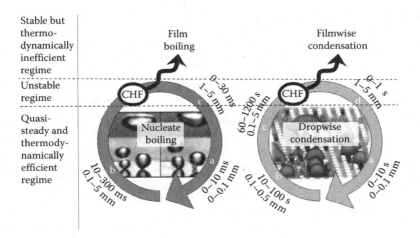

FIGURE 4.8 Synthetic representation and universal features of boiling and condensation heat transfer. (From Attinger, D., et al., *MRS Energ. Sustain. Rev. J.*, 1, E4, 2014. With permission.) The time and length scales are provided for each period.

TABLE 4.1

Major Forces and their Magnitude in Two-Phase Flow

			Magnitude of	
Forces	**Type**	**Force**	**Force per Unit Area**	**Force per Unit Volume**
Pressure force	Surface force	$A\Delta p$	Δp	$\Delta p/L$
Gravity, buoyancy	Volume force	$Vg\rho, Vg\Delta\rho$	$Lg\rho, Lg\Delta\rho$	$g\rho, g\Delta\rho$
Inertia	Volume force	$V\rho U^2/L$	ρU^2	$\rho U^2/L$
Viscous force	Surface force	$A\mu U/L$	$\mu U/L$	$\mu U/L^2$
Surface tension	Line force	$L\sigma$	σ/L	σ/L^2
Evaporation momentum	Surface force	$A(q/h_{lv})^2/\rho_v$	$(q/h_{lv})^2/\rho_v$	$(q/h_{lv})^2/(\rho_v L)$

are very different from those in conventional heat exchangers. For example, the surface tension force, which might be negligible in large channels, always plays an important role in microchannels. The gravity or buoyancy force, which affects the two-phase flow morphology in conventional channels, might be comparably small in microchannels.

4.3.3 COMPARISONS OF MULTIPHASE FLOW MODELING APPROACHES

Modeling of multiphase flow is very complex. The multiphase flow modeling approaches can be classified into three main categories: the Eulerian–Eulerian method (two-fluid model or multi-fluid model), the Eulerian–Lagrangian method, and direct numerical simulation (DNS, interface resolving method) [54,56,57]. Figure 4.9 presents the main characteristics of the three multiphase flow modeling approaches.

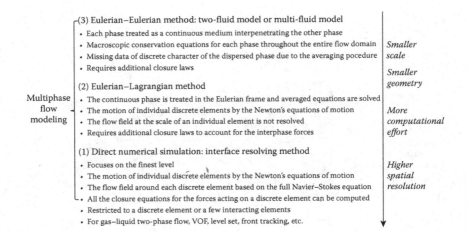

Multiphase flow modeling

(3) Eulerian–Eulerian method: two-fluid model or multi-fluid model
- Each phase treated as a continuous medium interpenetrating the other phase
- Macroscopic conservation equations for each phase throughout the entire flow domain
- Missing data of discrete character of the dispersed phase due to the averaging pocedure
- Requires additional closure laws

Smaller scale

Smaller geometry

(2) Eulerian–Lagrangian method
- The continuous phase is treated in the Eulerian frame and averaged equations are solved
- The motion of individual discrete elements by the Newton's equations of motion
- The flow field at the scale of an individual element is not resolved
- Requires additional closure laws to account for the interphase forces

More computational effort

(1) Direct numerical simulation: interface resolving method
- Focuses on the finest level
- The motion of individual discrete elements by the Newton's equations of motion
- The flow field around each discrete element based on the full Navier–Stokes equation
- All the closure equations for the forces acting on a discrete element can be computed
- Restricted to a discrete element or a few interacting elements
- For gas–liquid two-phase flow, VOF, level set, front tracking, etc.

Higher spatial resolution

FIGURE 4.9 Main characteristics of the three multiphase flow modeling approaches.

The Eulerian–Eulerian method can be applicable for industrial-scale simulations with comparatively less computational effort and lower spatial resolution compared to the other two methods. It seems that both the Eulerian–Eulerian method and the Eulerian–Lagrangian method are not applicable if one wants to capture the phase interface and its deformations in μHEXs, although DNS is a good choice. Only one set of governing equations are solved in DNS, which is also known as the one-fluid approach. DNS differs from the two-fluid and multi-fluid approaches from the point of view of the treatment of the phases in the computational domain. The former regards a multiphase flow as a single computational field in which more than one phase may exist, and the interface between two phases is solved as a part of the solution. In this kind of scheme, one set of governing equations suffices to calculate all essential flow field variables. In the light of difficulties in formulating the closure relations in the multi-fluid approach, especially in the transition regions of the flow pattern map, one-fluid schemes reveal their superiority. The method does not demand a priori shape of the interface geometry because this is part of the solution. This chapter will mainly focus on DNS for resolving the phase interface in microchannels and μHEXs.

4.3.4 Governing Equations for Gas–Liquid Flow

Considering incompressible Newtonian fluids for the phases *without phase change* in laminar flow in μHEXs, the transport of mass, momentum, and energy can be defined by the following set of partial differential equations [58]

$$\frac{\partial \rho}{\partial t} + \nabla \cdot (\rho u) = 0 \tag{4.10}$$

$$\frac{\partial (\rho u)}{\partial t} + \nabla \cdot (\rho u \cdot u) = -\nabla p + \nabla \cdot \left[\mu \left(\nabla u + \nabla u^{\mathrm{T}} \right) \right] + \rho g \tag{4.11}$$

$$\frac{\partial(\rho c_p T)}{\partial t} + \nabla \cdot (\rho c_p \boldsymbol{u} T) = \nabla \cdot (k \nabla T) + \psi \tag{4.12}$$

where ψ is the viscous dissipation.

When phase change occurs during two-phase flow, the governing differential equations of mass conservation, transport of momentum, and energy should consider the phase-change phenomenon appropriately. The mass conservation is rewritten as follows for incompressible Newtonian fluids [59]:

$$\nabla \cdot \boldsymbol{u} = \left(\frac{1}{\rho_v} - \frac{1}{\rho_l}\right) \dot{m} \delta(\boldsymbol{x} - \boldsymbol{x}_s) = \left(\frac{1}{\rho_v} - \frac{1}{\rho_l}\right) \dot{m} |\nabla \varepsilon| \tag{4.13}$$

where $\dot{m}$ is the mass flux intensity (per time per unit interfacial area) across the interface due to phase change, which will be explained later.

The single-fluid momentum conservation equation for Newtonian fluids in laminar flow takes the following form:

$$\frac{\partial(\rho \boldsymbol{u})}{\partial t} + \nabla \cdot (\rho \boldsymbol{u} \cdot \boldsymbol{u}) = -\nabla p + \nabla \cdot \left[\mu\left(\nabla \boldsymbol{u} + \nabla \boldsymbol{u}^{\mathrm{T}}\right)\right] + \rho \boldsymbol{g} + \boldsymbol{F}_\sigma \tag{4.14}$$

where $\boldsymbol{F}_\sigma$ is the surface tension force, which can be presented by the continuum surface force (CSF) method proposed by Brackbill et al. [60]:

$$\boldsymbol{F}_\sigma = \int_{\Gamma(t)} \sigma \kappa n \delta(\boldsymbol{x} - \boldsymbol{x}_s) ds = \sigma \kappa n |\nabla \varepsilon| \tag{4.15}$$

where $\Gamma(t)$ is the phase interface, $\delta(x - x_s)$ is a 3D Dirac delta function, n is the interface unit norm vector, κ is the local interface curvature, and ε is the volume fraction.

The energy equation is rewritten as

$$\frac{\partial(\rho c_p T)}{\partial t} + \nabla \cdot (\rho c_p \boldsymbol{u} T) = \nabla \cdot (k \nabla T) + \psi - \dot{m} \left[h_{lv} - \left(c_{p,v} - c_{p,l}\right)T\right] |\nabla \varepsilon| \tag{4.16}$$

The third term on the right-hand side shows the energy source terms given by the evaporation, the enthalpy of the created vapor, and the enthalpy of the removed liquid. The viscous dissipation term (i.e., the second term in Equation 4.16) can be safely neglected for water and refrigerants without loss of accuracy. The mass source term, the surface tension force, and the energy source terms are only present at the interface zone. Far from the phase interface, Equations 4.13, 4.14, and 4.16 become Equations 4.10, 4.11, and 4.12, respectively.

To complete the formulation, an equation for the interface temperature T_i needs to be provided. A simple assumption is to assume the interface temperature is equal to the equilibrium temperature corresponding to the system pressure

$$T_{\mathrm{sat}}\left(p_l\right) = T_l = T_i = T_v = T_{\mathrm{sat}}\left(p_v\right) \tag{4.17}$$

In many cases, this is an obvious and adequate assumption for macroscale boiling problems but, for microscale problems, Equation 4.17 cannot be assumed as a priori because the interfacial resistance to mass transfer across the interface should be considered. Using the kinetic theory [61,62], an approximate expression for the heat flux across a simplified liquid-vapor interface is written as

$$j_e^h = \alpha_e \left(T_i - T_{sat}\right) \tag{4.18}$$

where j_e^h is the evaporation heat-flux density. The evaporation heat transfer coefficient α_e is given by

$$\alpha_e = \frac{2\alpha}{2-\alpha} \cdot \frac{h_{lv}^2 \sqrt{M}}{T_{sat}^{3/2} \sqrt{2\pi R}} \cdot \frac{\rho_l \rho_v}{\rho_l - \rho_v} \tag{4.19}$$

where M is the molecular weight and R (8.314 J/mol) is the universal gas constant. Parameter α is the accommodation coefficient, which is also referred to as the evaporation or condensation coefficient. The accommodation coefficient is assumed to be equal for evaporation and condensation. Values of α vary widely in the literature [63]. For perfectly pure fluids, the accommodation coefficient is close to 1 for extremely pure fluids. Reported values of the accommodation coefficient range from 0.001 to 1. However, extreme purity is unlikely in most engineering systems. In order to quantify the deviations of the interfacial temperature from the vapor saturation temperature, let us consider a case of thin film evaporation, as shown in the inset of Figure 4.10. A liquid-vapor interface is located at a distance δ from a solid wall at constant temperature. From the energy balance

$$q_w = k_l \frac{T_w - T_i}{\delta} = j_e^h = \alpha_e \left(T_i - T_{sat}\right) \tag{4.20}$$

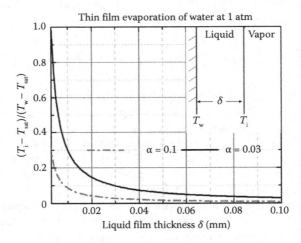

FIGURE 4.10 Dimensionless temperature difference versus film thickness for thin film evaporation of water at 1 atm.

One has

$$\frac{T_i - T_{sat}}{T_w - T_{sat}} = \frac{1}{1 + \alpha_e \delta / k_l} \tag{4.21}$$

The dimensionless temperature difference $(T_i - T_{sat})/(T_w - T_{sat})$ mainly depends on the fluid properties and the liquid film thickness δ. Figure 4.10 shows the relationship between the dimensionless temperature difference and the thin film thickness at different accommodation coefficients for water at 1 atm. The interfacial temperature jump $(T_i - T_{sat})$ might account for more than 10% of the wall superheat $(T_w - T_{sat})$ for very thin liquid films. Therefore, for numerical modeling of phase-change vapor-liquid flow in μHEXs, the simple assumption, Equation 4.17, is no longer valid. The interfacial temperature jump should be considered properly.

The mass flux intensity across the interface can be calculated by

$$\dot{m} = \frac{j_e^h}{h_{lv}} \tag{4.22}$$

The solution of the set of the governing equations consists of the flow variables such as pressure, velocity, and internal energy; however, the geometry of the phase interface is not provided. Values of the time- and space-dependent physical properties of the flow cannot be assessed. In DNS such as the volume of fluid (VOF) method and the level set (LS) method, a scalar quantity is introduced to indicate the position of the phase interface. This term is a scalar field representing the phase distribution. In general, the quantity, say $\phi(x, t)$, must satisfy the following transport equation for adiabatic gas–liquid two-phase flow:

$$\frac{\partial \phi}{\partial t} + u \cdot \nabla \phi = 0 \tag{4.23}$$

Distribution of $\phi(x, t)$ in space can be derived by the aforementioned transport equation. In the VOF method, the volume of fraction function $\phi(x, t)$ has a value between 0 and 1. The volume of fraction is one when the cell is full of the primary phase, zero if the cell is full of the secondary phase, and $0 < \phi < 1$ when both phases present and the phase interface exists in the cell. Volume fraction ε represents the ratio of the cell volume occupied by the primary phase, which can be obtained from $\phi(x, t)$. For example, the volume fraction of the VOF method is calculated by integration of $\phi(x, t)$ over the computational cell of volume V:

$$\varepsilon = \frac{1}{V} \int_V \phi(x, t) dV \tag{4.24}$$

In the level set method, the scalar quantity $\phi(x, t)$ is a level set function that represents the interface among immiscible fluids. The level set function represents the distance to the interface and is equal to zero on the interface. Positive and negative ϕ values indicate the two immiscible phases. The physical properties of the mixture for every domain cell, such as density, specific heat capacity, viscosity, and thermal conductivity, are updated locally based on $\phi(x, t)$. The properties are usually the weight-averaging values of all the phases in the cell and have the following form

$$\Phi = \sum_{1}^{n} (\Phi_i \varepsilon_i) \qquad (4.25)$$

where Φ represents a relevant physical property.

To evolve the interface location for gas–liquid two-phase flow with phase change, the following volume fraction or level set conservation equation needs to be solved instead

$$\frac{\partial \phi}{\partial t} + \nabla \cdot (\phi u) = \frac{1}{\rho_v} \dot{m} |\nabla \phi| \qquad (4.26)$$

4.3.5 Interface Resolving Methods

Various ideas have been proposed for interface evolution. There are typically two groups: interface capturing and interface tracking. The interface capturing method uses advection of a transport function for an implicit description of the interface on a fixed-grid computational domain, as shown in Figure 4.11a. The most common interface capturing methods are the level set method and the VOF method. In comparison, the interface tracking method is based on a moving-grid domain. It simulates the interface behavior by deforming the elements to represent the interface. The interface is explicitly described by the computational mesh, as seen in Figure 4.11b. A commonly used interface tracking method is the arbitrary Lagrangian and Eulerian (ALE) method. Table 4.2 summarizes the advantages and limitations for several interface resolving methods (i.e., VOF, the level set, phase field, front-tracking, and moving mesh methods). All these methods are continuum methods. The coupling of the interface evolution equation with the single-field Navier–Stokes equations and related issues such as surface tension forces and contact lines were presented in the comprehensive review by Wörner [29]. Therefore, only a brief description of the common continuum methods such as VOF, the level set, and the front-tracking methods will be outlined here.

The VOF method defines a volume of fluid function (or the so-called color function or the smoothed color function) that represents the phase fraction. The interface evolution of the VOF method is governed by transport equations similar to Equations 4.23 and 4.26 for adiabatic and phase-change cases, respectively. The fluid interface is located where the volume fraction function shows a sharp transition between the two values representing the two different phases [65,66]. The VOF scheme was developed by Hirt and Nichols [67]. Welch and Wilson [58] implemented a phase-change scheme in the VOF method based on Young's enhancement [68] for discontinuous interfacial properties. A drawback of the VOF method is the existence of spurious currents. To calculate the curvature as accurately as possible, a height function approach [59] or the parabolic reconstruction of surface tension (PROST) [69] must

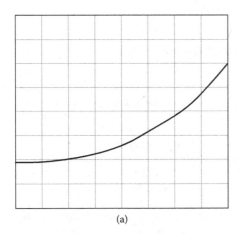
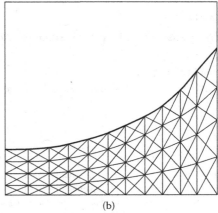

(a)   (b)

FIGURE 4.11 Interface resolving methods: (a) the interface capturing method and (b) the interface tracking method. (From Okamori, K., *Free surface flow analysis,* available at: http://www.cradle-cfd.com/tec/column02/002.html.)

be implemented to effectively eliminate spurious currents. Recently, the VOF scheme has been extended to model phase-change two-phase flow in microchannels [70–74].

The level set approach is similar to the VOF scheme in the sense that a scalar field quantity—the level set function—signaling the presence of either one of the two phases is advected on a fixed grid. The most common level set function is the distance function. In this case, each cell has a distance function with positive or negative signs that indicate two different phases, and the magnitude is equal to the distance to the closest interface. Thus, the zero level set contour defines the interface location. Sussman et al. [75] developed a level set approach where the interface was captured implicitly as the zero level set of a smooth function. Son et al. [76] modified the level set method to include the liquid-vapor phase change and simulated the growth and departure of single bubbles during nucleate pool boiling. Mukherjee and Dhir [77] extended the model to 3D cases and studied the merger and departure of multiple bubbles during pool boiling. Later, the 3D level set method was adopted by Mukherjee and Kandlikar [78] and Mukherjee [79] to simulate the bubble dynamics during flow boiling of water in a microchannel. Suh et al. [80] studied the bubble dynamics and the associated flow and heat transfer in parallel microchannels by means of a level set method, in order to investigate the conditions leading to flow reversal. Numerically, the level set method is not inherently conservative. The problem has been addressed in previous level set implementations [81–83] through the addition of a volume preservation constraint, that is, the coupled level set and volume of fluid method (CLSVOF). In CLSVOF, the level set function is used to compute the interface curvature normal to the interface, while the VOF function is used to capture the interface. A detailed description of the CLSVOF method was given by Nichita [83]. Figure 4.12a shows the bubble contour with both level set and VOF functions when solving the coupling equation after 0.2 s. In Figure 4.12b, the same contours are shown without the coupling between LS and VOF. It is clear that: (i) the LS

TABLE 4.2

Comparison of Several Interface Resolving Methods

Interface Evolution Methods	Advantages	Limitations/Challenges
VOF	• Inherent mass conservation • Relatively simple and accurate • Can detect topological changes • Lots of quantitative validations with experimental data	• Complex interface reconstruction • The conservation of energy cannot be maintained during computation • Difficult to obtain the accurate curvature and to smooth the discontinuous physical quantities
Level set	• Conceptually simple and easy to implement • Easy to compute derivatives of the color function • Continuous interface and no need for complex interface reconstruction • Handle topological changes and complex interfacial shapes in a simplified manner • Many codes can be converted from 2D to 3D easily	• Nonconservation of mass • The signed distance function must be reinitialized after each time step • The conservation of energy cannot be maintained during computation • Limited accuracy
Phase field	• Energy conservation • Straightforward numerical solution of a few equations • Simplify the handling of topological changes • Can simulate contact line motion • Locations of the interfaces no longer need to be tracked but can be inferred from the field parameters	• Mass loss • Large domains computationally challenging • Interface width is an adjustable parameter which may be set to physically unrealistic values • Ambiguity in choosing the dynamics of the phase field and the parameter values
Front-tracking	• Accurately represent the interface as continuous surface • Extremely accurate and robust • Provide subgrid phase interface resolution	• Troublesome interface reconstruction • Dynamic remeshing required • Difficult to track topological changes • Computationally expensive • Issues of numerical instabilities • Complex in 3D
Moving mesh	• Accurately represent the interface with fast convergence • Interpolation of dependent variables from the old mesh to the new mesh is unnecessary • Automatically detect, resolve, and track steep wave fronts and moving boundaries • Can work for problems with large gradients or discontinuities	• Does not support topological changes • Large domains computationally challenging • Increases the stiffness of the system and requires implicit time integration to work efficiently • Choosing time scale in equidistributing moving mesh methods is highly heuristic

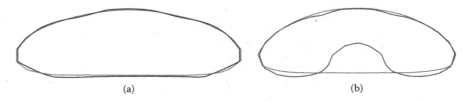

(a)   (b)

FIGURE 4.12 Level set contour (black) and volume of fluid (VOF) contour (gray) (a) after solving the coupled equation between level set and VOF and (b) without coupling between level set and VOF. (From Nichita, B.A., *An Improved CFD Tool to Simulate Adiabatic and Diabatic Two-Phase Flows*, PhD thesis, EPFL, Lausanne, Switzerland, 2010. With permission.)

method lost a significant fraction of mass and (ii) by coupling LS with VOF, one can conserve mass in case of an LS method. One disadvantage of the CLSVOF method is that local volume correction might introduce large errors in higher derivatives of the level set function. Another method to address the mass nonconservation is the refined level set grid method, which increases accuracy of the level set solution by increasing grid resolution. Herrmann [84] developed an efficient refined level set grid method to ensure good fluid volume conservation properties. The refined level set grid method is illustrated in Figure 4.13.

A front-tracking method based on DNS was introduced by Unverdi and Tryggvason [85]. The phase interface is tracked by Lagrangian marker points in a fixed grid. A marker point lying on the interface at position x_p is advected by the flow according to

$$\frac{d\boldsymbol{x}_p}{dt} = \boldsymbol{u}_p \tag{4.27}$$

The velocity $\boldsymbol{u}_p$ at position $\boldsymbol{x}_p$ is determined from the velocity field on the Eulerian grid by interpolation. In order to keep the interface adequately resolved throughout the simulation, a remeshing procedure is performed where marker points may be added or removed. Contrary to other numerical models, such as the VOF and level set methods, the front-tracking method uses an unstructured dynamic mesh to represent the interface surface and tracks this interface explicitly by the interconnected marker points [85–87]. The Lagrangian representation of the interface avoids the necessity of reconstructing the interface from the local distribution of the fractions of the phases and, moreover, allows a direct calculation of the surface tension forces without the inaccurate numerical computation of the interface curvature [87]. As compared with the VOF and level set techniques, which are ubiquitous in the multiphase CFD community, front-tracking codes have been developed by only a few groups [85–89] so far, though their number is increasing [29].

The aforementioned interface resolving approaches are continuum methods. In recent years, the lattice Boltzmann method (LBM), as a mesoscopic numerical method, has shown great potential in modeling multiphase fluid flows (e.g., He and Doolen [90], Inamuro et al. [91], Zheng et al. [92], Gong et al. [93], and Gong

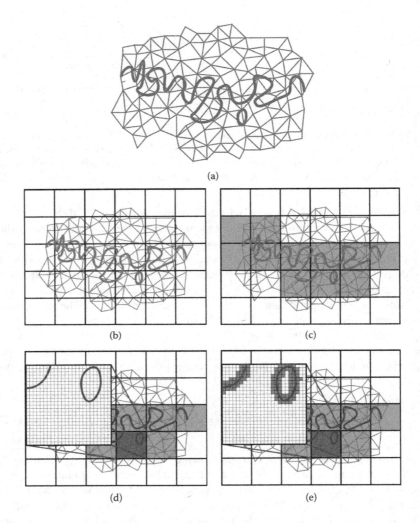

FIGURE 4.13 The refined level set grid method: (a) front on a flow-solver grid, (b) intro-duce equidistant Cartesian super-grid (blocks), (c) activate (store) only narrow band of blocks (gray), (d) activate the dark blocks consist of an equidistant Cartesian fine G-grid, and (e) acti-vate (store) only narrow band of fine G-grid. (From M. Herrmann, *Annual Research Briefs*, Center for Turbulence Research, 2005. With permission.)

and Cheng [94]). Generally, the interface is tracked by an index function and has a thickness of several cell sizes. Interfaces are implicitly defined by the fluid frac-tion isosurface where the content of the two fluids is equal. The surface tension can be modeled in the LBM by incorporating molecular interactions. Because the nonlinear Navier–Stokes equations in the continuum methods are replaced by the semilinear Boltzmann equation, the LBM is easier to be implemented than con-tinuum methods. For multiphase flows, the LBM does not track interfaces because phase separation occurs automatically. Some additional advantages of the LBM are ease of parallelization and the simplicity in dealing with domains with complex geometries. The kinetic nature of the LBM provides many of the advantages of molecular dynamics, making the LBM particularly useful in simulating complex interfacial dynamics. On the other hand, the LBM has some disadvantages over conventional continuum methods: (a) macroscopic fluid properties and transport coefficients cannot be prescribed as input parameters; (b) the LBM method is difficult to generalize beyond its domain of validity, and the jump conditions across the interface are difficult to satisfy; (c) lack of strict mass conservation and substantial additional CPU time for global mass correction procedures; and (d) limited to relatively small values of density and viscosity ratios for the reason of numerical stability. Recently, Cheng and his colleagues [93–95] developed an LBM to simulate phase-change two-phase flows with liquid-vapor density ratios up to 1000.

4.3.6 CFD APPLICATIONS FOR PHASE-CHANGE TWO-PHASE FLOW IN MICROCHANNELS

There is a scarcity of numerical work in the literature for two-phase flow in μHEXs. Because microchannels are the basic elements of μHEXs, recent CFD applications of two-phase flow (phase-change involved) in microchannels, mostly limited to the scale of a few bubbles or droplets, are briefly summarized and discussed in this section. Table 4.3 summarizes previous literature studies on numerical simulation of phase-change two-phase flow in microchannels. This section will mainly concern numerical modeling of phase-change two-phase flow. For numerical modeling of adiabatic gas–liquid two-phase flow or liquid-liquid two-phase flow in microchannels, interested readers may refer to the comprehensive review by Talimi et al. [96]. As noted in Table 4.3, bubble dynamics was covered in most of the flow boiling simulations, while condensation studies focused on annular film condensation. A 2D VOF model was proposed by Ganapathy et al. [72] to simulate R134a condensation heat transfer and fluid flow in a single microchannel with a characteristic dimension of 100 μm. Mass transfer at the liquid-vapor interface and the associated release of latent heat were included in the governing equations. The VOF approach was implemented by a finite volume method. The surface tension forces were modeled by the continuum surface force formulation of Brackbill et al. [60]. The pressure jump across the interface was modeled as a volume force that was included as a source term in the momentum equation. Interface reconstruction was performed using the explicit piecewise-linear interface reconstruction (PLIC) scheme of Youngs [68]. Figure 4.14 shows the comparisons between the predicted flow patterns (regimes) and the experimental visualization data from the literature for condensing flows. Because the channel geometry and operating conditions for the simulation do not correspond to the experimental channel geometry and working conditions, the flow regime comparison is qualitative in nature. As shown in Figure 4.14, the numerical model has a high qualitative accuracy in predicting flow regimes. Besides, the quantitative predictive accuracy of the numerical model was assessed by comparing the two-phase frictional pressure drop and Nusselt number against numerous empirical correlations in the literature. However, because existing empirical correlations in

TABLE 4.3

Summary of Selected Literature Work on Numerical Simulation of Two-Phase Flow (Phase-Change Mostly Involved) in Microchannels

Authors	Heat Transfer Phenomena	Multiphase Flow Model	Experimental Validation
Mukherjee and Kandlikar [78], 2005	Single bubble dynamics during flow boiling water	Level set, 3D	Transient bubble profiles
Chung et al. [97], 2009	Droplet breakup and merger past a cylinder obstruction, Newtonian and viscoelastic fluids	Front tracking, 2D	Droplet dynamics
Fang et al. [98], 2010	Flow boiling in a vapor-venting microchannel, water	VOF, 3D	Flow regimes; liquid rise velocity in porous medium
Nebuloni and Thome [99], 2010	Bubble dynamics during flow boiling, water, R134a	Coupled level set and VOF (CLSVOF)	Transient bubble profiles, liquid film thickness
Wang et al. [100], 2011	Droplet formation in Venturi-shaped microchannels	LBM, 2D, and 3D	Droplet dynamics in a T-shaped microchannel
Da Riva and Del Col [73], 2012	Annular laminar film condensation, R134a	VOF, 3D	Evolution of the interface, heat transfer coefficient
Lee et al. [101], 2012	Flow boiling in a finned microchannel, water	Level set, 3D	Bubble dynamics, heat transfer
Nebuloni and Thome [102], 2012	Conjugate heat transfer for annular laminar film condensation, R134a	Theoretical and empirical correlations	Heat transfer coefficient
Zhuan and Wang [71], 2012	Flow boiling regimes, R134a	VOF, 3D	Bubble departure diameter, location, and frequency; flow regimes
Ganapathy et al. [72], 2013	Convective condensation, R134a	VOF, 2D	Flow regimes, pressure drop, and heat transfer
Magnini et al. [59, 103], 2013	Single and multiple bubbles on slug flow boiling, various fluids	VOF, 2D	Liquid film thickness, bubble shape, bubble nose position
Miner and Phelan [104], 2013	Flow boiling in expanding microchannels, various fluids	Theoretical and empirical correlations	Critical heat flux
Chen et al. [105], 2014	Convective condensation, FC72	VOF, 3D	Flow regimes
Gong and Cheng [95], 2014	Saturated flow boiling, water	LBM, 2D	Isothermal single-phase velocity profile
Lan et al. [106], 2014	Droplet formation	Level set, 3D	Droplet shape and size
Yin et al. [74], 2015	Laminar film condensation, water	VOF, 3D	Annular flow

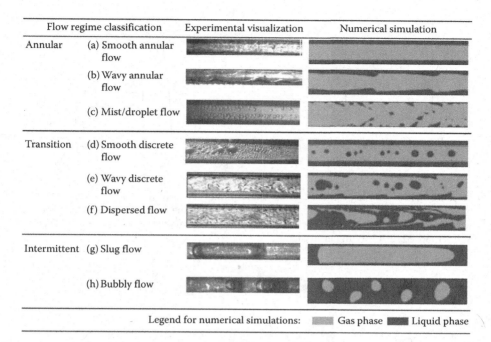

Flow regime classification		Experimental visualization	Numerical simulation
Annular	(a) Smooth annular flow		
	(b) Wavy annular flow		
	(c) Mist/droplet flow		
Transition	(d) Smooth discrete flow		
	(e) Wavy discrete flow		
	(f) Dispersed flow		
Intermittent	(g) Slug flow		
	(h) Bubbly flow		

Legend for numerical simulations: ▨ Gas phase ▨ Liquid phase

FIGURE 4.14 Comparison of numerically predicted condensation flow regimes with experimental visualization data in the literature. (From Ganapathy, H., et al., *Int. J. Heat Mass Tran.*, 65, 62–72, 2013.) Experimental visualization images. (From Kim, S.M., et al., *Int. J. Heat Mass Tran.*, 55, 971–983, 2012; Wu, H.Y., et al., *Int. J. Heat Mass Tran.*, 48, 2186–2197, 2005; Hu, J.S., et al., *Int. J. Refri.*, 30, 1309–1318, 2007; Coleman, U.W., et al. *Int. J. Refri.*, 26, 117–128, 2003. With permission.)

microchannels often have limited applicable range and large deviations from each other, they might not be accurate enough for validation.

Magnini et al. [59] simulated single elongated bubbles during flow boiling in a circular microchannel in the commercial CFD code ANSYS Fluent 12 with a VOF method for interface capturing. A height function algorithm was implemented to replace the ANSYS Fluent default method (the Young's PLIC formulation) to estimate the local curvature. Interested readers may refer to Magnini et al. [59] for details of the height function interface reconstruction algorithm. The numerical method was extended to simulate the influence of leading and sequential bubbles on slug flow boiling within a microchannel by the same authors [103]. A circular microchannel with a diameter of 0.5 mm was modeled as a two-dimensional axisymmetric channel with an initial adiabatic region followed by a heated region. The length of the adiabatic region was chosen to ensure that the bubbles enter the heated region in a steady-state flow condition. Figure 4.15 shows the profiles of the bubble in the initial adiabatic region, the leading bubble and the trailing bubble in the heated section. The profiles were shifted in order to match the nose positions for comparison. The leading bubble (bubble ahead) has grown to a length of 9D from 3D of the bubble in the adiabatic region. The bubble nose is less blunt than the adiabatic profile, and the liquid film of the leading bubble

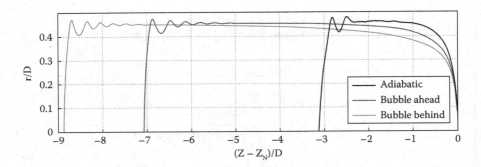

FIGURE 4.15 Bubble profiles: The adiabatic profile refers to that of the leading bubble before it enters in the heated region. The profiles of the leading bubble and the trailing bubble were captured after 19 ms and 31 ms, respectively. (From Magnini, M., et al., *Int. J. Therm. Sci.*, 71, 36–52, 2013. With permission.)

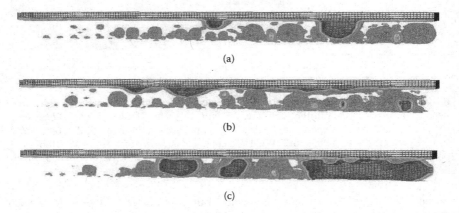

FIGURE 4.16 Impact of membrane surface wettability on the vapor-venting flow pattern at a heat flux of 10 W/cm². (a) Moderately hydrophobic membrane surface (114.5°); (b) hydrophobic membrane surface (179°); and hydrophilic membrane surface (28.3°). (From Fang, C., et al., *Front. Heat Mass Tran.*, 1, 013002, 2010. With permission.)

profile is thicker than that of the adiabatic profile. The trailing bubble is noticeably shorter (7D) than the leading one due to lower growth rate. This in turn gives rise to less bubble acceleration and hence a more rounded profile of the bubble nose and a slightly thinner liquid film than the leading bubble. It seems that the heat transfer coefficient of the trailing bubble is higher than that of the leading bubble.

Fang et al. [98] numerically simulated the 3D transient vapor-venting process in a rectangular microchannel bounded on one side by a hydrophobic porous membrane for phase separation of a vapor-venting microchannel heat exchanger using the VOF method along with models for interphase mass transfer and capillary force. Impact of membrane surface wettability on the vapor-venting flow pattern is shown in Figure 4.16. The bubbly flow (Figure 4.16a) and the elongated slug flow (Figure 4.16c) were observed in microchannels with moderately hydrophobic membrane and hydrophilic membrane, respectively. In contrast, the bubbles in the superhydrophobic case (Figure 4.16b) tend to spread on the membrane surface and form a vapor buffer layer, which significantly increases the contact area between the vapor and the membrane, therefore enhancing the vapor-venting efficiency.

Flow boiling of a vapor bubble in a square microchannel was systematically studied by Mukherjee and his colleagues [77–79] using a level set method to track the interface. For example, Figure 4.17 compares the bubble shapes between the numerical simulation and experimental data. Figure 4.17a shows the numerically obtained bubble shapes by the level set method. Figure 4.17b shows the experimental data at similar times with the bubble outline indicated in each frame. The bubbles are seen to move downstream along with the flow and to touch the side walls around 1.0 ms. Vapor patches can be observed on the surface between the bubble end caps in the experimental data at 1.4 and 1.8 ms. Similar vapor patch formation on the wall is observed in the numerical simulation at 1.39 and 1.8 ms. Comparison of bubble growth rate and shapes indicate good agreement between the numerical results and the literature experimental data and provides validation for the numerical model.

The level set method can simulate the bubble dynamics in enhanced microchannels. Direct numerical simulations of flow boiling in a microchannel with transverse

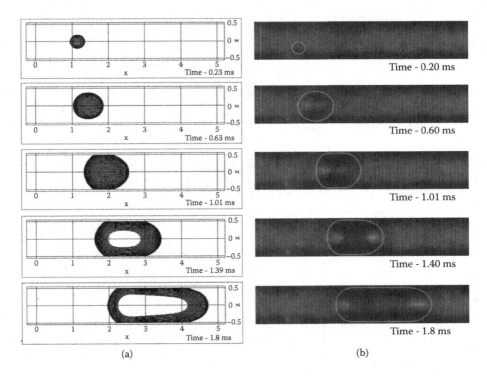

(a)   (b)

FIGURE 4.17 Comparison of bubble shapes: (a) numerical simulation, (b) experimental data. (From Mukherjee, A., et al., *Int. J. Heat Mass Tran.*, 54, 3702–3718, 2011. With permission.)

fins were performed by using a sharp-interface level set method by Lee et al. [101]. The computational results showed that the flow boiling in a microchannel was significantly enhanced when the liquid–vapor–solid interface contact region increased in finned microchannels. In Figure 4.18, the liquid layers between the fins are thin near the channel center ($z = 0$) and become thicker near the channel sidewalls. Compared to smooth microchannels, the liquid–vapor–solid contact regions also increase on the sidewalls, as seen in the cross-sections of $x = 3.05$ (fin region) and $x = 3.15$ (interfin region). As liquid is trapped between the fins, the liquid-vapor interface contacts the finned surface at more locations and hence enhances the heat transfer rates.

Chung et al. [97] investigated the droplet dynamics passing through a cylinder obstruction in a confined microchannel flow with DNS by a finite element front-tracking method. A single droplet might show manifold deformations when passing through the cylinder obstruction. The transient dynamics of a droplet with a length 1.3 times that of the channel width was present in Figure 4.19 at a capillary number of 0.05. In Figure 4.19a, the droplet shows a bullet shape in the confined microchannel, and it splits approaching the cylinder obstruction, as shown in Figure 4.19b. The width of a thread rounding the cylinder obstruction becomes thinner as in Figure 4.19c, and the thread breaks up, as shown in Figure 4.19d, generating

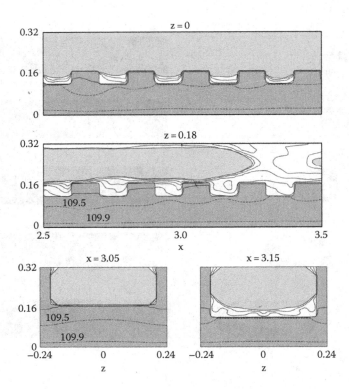

FIGURE 4.18 Temperature field at $t = 3$ ms in a finned microchannel. (From Lee, W., et al., *Int. Commun. Heat Mass Tran.*, 39, 1460–1466, 2012. With permission.)

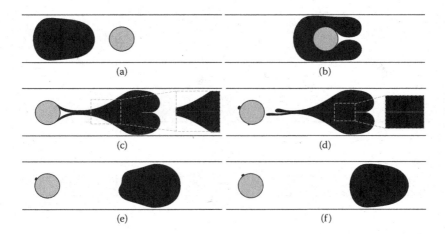

FIGURE 4.19 Transient dynamics of a short droplet passing through the cylinder obstruction. (a) Upcoming bullet shape at $t = 3.6$, (b) droplet split at $t = 5.6$, (c) thread thinning at $t = 8.4$, (d) thread breakup at $t = 8.8$, (e) merger at $t = 9.2$, and (f) final shape at $t = 10$. (From Chung, C., et al., *J. Non Newtonian Fluid Mech.*, 162, 38–44, 2009. With permission.)

satellite droplets. If two droplets flow closely, a merger into a single droplet occurs, as shown in Figure 4.19e, and the droplet shows a recovery process into the steady bullet shape again as in Figure 4.19f. The time (t) in Figure 4.19 was nondimensionalized with a characteristic time scale 0.5 w/U, where w and U are the channel width and the mean inlet velocity, respectively.

The LBM has shown great potential in modeling complex fluid systems. In the LBM, the fluid is described by the evolution of microscopic fluid particles or particle distribution functions. The LBM is based on mesoscopic kinetic equations and can incorporate interparticle interactions directly. Recently, Gong and Cheng [94] have studied the force term and the method of incorporating the force term individually in the single-component multiphase LBM and found that both of them play an important role in the performance for numerical simulation of multiphase flows. The proposed model was applied to simulate three-dimensional droplet motion and coalescence processes on bottom surfaces with wettability gradients, as shown in Figure 4.20 [111]. Figure 4.20a shows the three-dimensional coalescence process of two droplets driven by wettability discontinuities from $t = 10,520$ to $t = 10,820$. Figure 4.20b is a top view showing the changing shapes of the droplets during the same period of time. The long axis of the droplet is in x-direction at $t = 10,660$ and in y-direction at $t = 10,820$, indicating that the shape of the droplet is experiencing an oscillation process after coalescence and before the final equilibrium state is reached.

Gong and Cheng [95] simulated saturated flow boiling in microchannels at low Reynolds numbers by a two-dimensional LBM. Two particle distribution functions, namely the density distribution function and the temperature distribution function, were used in this model. A new form of the source term in the energy equation was derived, and the modified pseudo-potential model was used in the proposed model to improve its numerical stability. The commonly used Peng–Robinson equation of

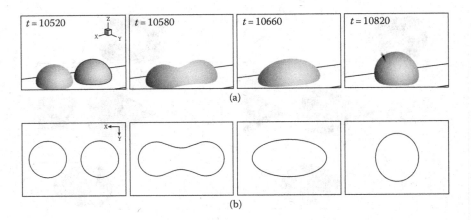

(a)

(b)

FIGURE 4.20 Coalescence of two droplets on the bottom surface with wettability gradients in a microchannel. (a) 3D droplet(s) shape and (b) top view of the droplet(s). (From Gong, S. and Cheng, P., *Comput. Fluid.*, 53, 93–104, 2012. With permission.)

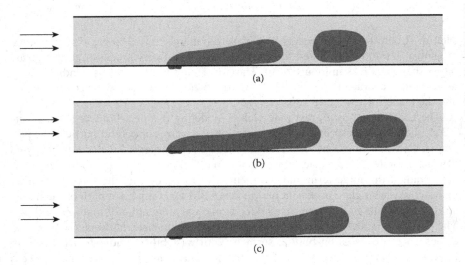

(a)

(b)

(c)

FIGURE 4.21 Effect of contact angle on bubble formation in a microchannel under constant heat-flux condition with $Re_l = 0.45$ and $q = 0.01$ at $t = 72,000$ for three static contact angles: (a) $\theta = 57°$; (b) $\theta = 96°$; and (c) $\theta = 118°$. (From Gong, S., and Cheng, P., *Numer. Heat Tran. A Appl.*, 65, 644–661, 2014. With permission.)

state was incorporated into the proposed model. Effect of contact angle on bubble formation process for three static contact angles was simulated using the proposed model, as shown in Figure 4.21. The departed bubble size is almost the same for the three contact angles, indicating that the departed bubble size is independent of the static contact angle in flow boiling in microchannels. As the nucleation time is affected by the contact angle, the bubble front at the same time step appears at different positions on the channel wall. The results of these applications suggest that the LBM model is an effective tool for simulation of phase transitions and multiphase flows and has the potential of simulating more complex fluid problems at a wider temperature range [95].

4.4 CONCLUSIONS AND FUTURE RESEARCH NEEDS

A state-of-the-art review has been carried out on numerical modeling of single-phase flow and two-phase flow in µHEXs and microchannels, and relevant selected CFD applications have been presented briefly in this chapter. Governing equations were given mainly for the laminar flow. The main concluding remarks and relevant future research needs are stated as follows.

For numerical modeling of single-phase flow

- Entrance effects, conjugate heat transfer, viscous dissipation, thermophysical property variations, inlet and outlet plenums, and flow distribution need to be appropriately incorporated into the numerical models of µHEXs, depending on the specific µHEX applications.
- Scaling effects, often negligible in conventional channels, might be significant in microchannels. The two dimensionless numbers given in Equations 4.5 and 4.9 might be used to evaluate the importance of conjugate heat transfer and viscous dissipation on fluid flow and heat transfer in µHEXs, respectively.
- The governing equations in the µHEXs were oversimplified in many cases. For example, thermophysical property variations were neglected in many numerical works [42,50]. Conjugate heat transfer was also not considered appropriately in some numerical studies.
- Most previous numerical studies were focused on one microchannel or a part of the microchannel array. Fluid dynamics and heat transfer in the microchannel can be simulated in full detail. However, the results from a part of the µHEX might not be extended to the whole µHEX because the effects of inlet and outlet plenums and flow maldistribution are not properly addressed.
- Numerical simulations can aid development of efficient µHEXs by coupling with constructal theory or other optimization procedures.
- No systematic experiments are available to validate the numerical models quantitatively.

For numerical modeling of two-phase flow

- Presented the challenges of modeling two-phase flow and the significant forces in two-phase flow.
- Characteristics of the multiphase flow modeling approaches (i.e., the Eulerian–Eulerian method, the Eulerian–Lagrangian method, and DNS) were compared. The Eulerian–Eulerian method can be applicable for industrial-scale simulations with comparatively less computational effort and lower spatial resolution compared to the other two methods. DNS (interface resolving method) is a good choice to capture/track the phase interface and its deformations in µHEXs.

- Governing equations for adiabatic two-phase flow and phase-change two-phase flow were presented. Only one set of governing equations was solved in the DNS. Interface mass transfer and source terms were given for phase-change two-phase flow.
- Advantages and disadvantages of several continuum interface resolving methods, such as VOF, level set, phase field, front-tracking, and moving mesh methods, were outlined. The characteristics of the LBM were also discussed. The LBM, as a mesoscopic numerical method, has shown potential in modeling phase-change two-phase flow.
- Numerically, the level set method is not inherently conservative. The problem can be addressed by implementing the level set method through the addition of a volume preservation constraint (e.g., the coupled level set and VOF method) or by increasing grid resolution (e.g., the refined level set grid method) to ensure good fluid volume conservation properties.
- CFD applications of two-phase flow (phase-change involved) in microchannels were briefly summarized and presented, mostly limited to the scale of a few bubbles or droplets. The numerical models were more or less validated qualitatively by comparing the interface shapes and flow regimes among the numerical simulation and experiments. For the VOF method, many quantitative validations with experimental data or empirical correlations were also performed.
- Because the channels in microfluidic devices and μHEXs are square or rectangular, more three-dimensional computations are required.
- Knowledge about effects of surface characteristics such as surface roughness and wettability on the phase-change two-phase flow needs to be obtained in the future.
- Suitable detailed and systematical experimental databases are required for quantitative and sound validation of these numerical methods and models in the future, especially for phase-change two-phase flow.
- Although interface resolving methods show some promising results in solving simple problems, applications to complex engineering problems (in μHEXs) much more investigation is still needed.
- Robust multiscale methods need to be developed in the future that combine models valid at different length scales and account for their interaction.

Although there are many hurdles to overcome, continued research on numerical modeling will help us to understand the transport phenomena in complex microscale geometries and therefore to develop next-generation, ultra-compact, and efficient μHEXs in the future.

NOMENCLATURE

A	Area
Bo	Bond number
Br	Brinkman number
c_p	Specific heat

d	Diameter
d_h	Hydraulic diameter
F_σ	Surface tension force
g	Gravitational acceleration
h	Heat transfer coefficient
h_{lv}	Latent heat
j_e^h	Evaporation heat-flux density
k	Thermal conductivity
Kn	Knudsen number
L	Characteristic dimension; length
M	Maranzana number; molecular weight
$\dot{m}$	Mass flux intensity across the interface
n	Interface unit norm vector
Nu	Nusselt number
p	Pressure
Pr	Prandtl number
q	Heat flux
R	Universal gas constant
r	Radial coordinate
Re	Reynolds number
Re$_l$	Liquid Reynolds number
S	Source term
T	Temperature
t	Time
U	Mean velocity
u	Velocity vector
u_p	Particle velocity
V	Volume
x	Distance along the x-coordinate
x_p	Particle position

Greek Symbols

Δp	Pressure drop
α	Accommodation coefficient
α_e	Evaporation heat transfer coefficient at the interface
δ	Liquid film thickness
ε	Volume fraction
θ	Contact angle
κ	Local interface curvature
λ	Molecular mean free path
μ	Dynamic viscosity
ρ	Density
σ	Surface tension
φ	Arbitrary dependent variable in Equation 4.2
ψ	Viscous dissipation
ϕ	Scalar quantity

| Γ | Generalized diffusion coefficient |
| Φ | Relevant physical property |

Subscripts

f	Fluid
i	Interface
l	Liquid
s	Solid
sat	Saturated
v	Vapor
w	Wall

ACKNOWLEDGMENTS

We gratefully acknowledge financial support from the Swedish National Research Council and the Swedish Energy Agency. The cooperation between University of Rhode Island, Kingston, and Lund University was partly supported by the Wenner-Gren Foundation.

REFERENCES

1. T. Dixit and I. Ghosh, Review of Micro- and Mini-Channel Heat Sinks and Heat Exchangers for Single Phase Fluids, *Renewable and Sustainable Energy Reviews*, vol. 41, pp. 1298–1311, 2015.
2. D.B. Tuckerman and R.F.W. Pease, High-Performance Heat Sinking for VLSI, *IEEE Electron Device Letters*, vol. 2, pp. 126–129, 1981.
3. Z. Wu and B. Sundén, On Further Enhancement of Single-Phase and Flow Boiling Heat Transfer in Micro/Minichannels, *Renewable and Sustainable Energy Reviews*, vol. 40, pp. 11–27, 2014.
4. D. Reay, C. Ramshaw, and A. Harvey, *Process Intensification: Engineering for Efficiency, Sustainability and Flexibility*, Butterworth-Heinemann, Oxford, 2013.
5. X. Yin, *Micro Heat Exchangers*, PhD dissertation, University of Pennsylvania, Philadelphia, PA, 1995.
6. B.G. Carman, J.S. Kapat, L.C. Chow, and L. An, Impact of a Ceramic Microchannel Heat Exchanger on a Micro Turbine, *ASME Turbo Expo 2002: Power for Land, Sea, and Air*, pp. 1053–1060, 2002.
7. H.H. Bau, Optimization of Conduits' Shape in Micro Heat Exchangers, *International Journal of Heat and Mass Transfer*, vol. 41, pp. 2717–2723, 1998.
8. Y. Peles, A. Koşar, C. Mishra, C.J. Kuo, and B. Schneider, Forced Convective Heat Transfer across a Pin Fin Micro Heat Sink, *International Journal of Heat and Mass Transfer*, vol. 48, pp. 3615–3627, 2005.
9. P.A. Kew and D.A. Reay, Compact/Micro-Heat Exchangers—Their Role in Heat Pumping Equipment, *Applied Thermal Engineering*, vol. 31, pp. 594–601, 2011.
10. B. Sundén and Z. Wu, Advanced Heat Exchangers for Clean and Sustainable Technology, in *Handbook of Clean Energy Systems*, ed. J. Yan, Wiley, West Sussex, UK, 2015.
11. B. Alm, U. Imke, R. Knitter, U. Schygulla, and S. Zimmermann, Testing and Simulation of Ceramic Micro Heat Exchangers, *Chemical Engineering Journal*, vol. 135, pp. S179–S184, 2008.

12. S.K. Mylavarapu, X. Sun, R.N. Christensen, R.R. Unocic, R.E. Glosup, and M.W. Patterson, Fabrication and Design Aspects of High-Temperature Compact Diffusion Bonded Heat Exchangers, *Nuclear Engineering and Design*, vol. 249, pp. 49–56, 2012.
13. R. Sayegh, M. Faghri, Y. Asako, and B. Sunden, *Direct Simulation Monte Carlo of Gaseous Flow in Microchannel, presented at 33rd National Heat Transfer Conference*, Albuquerque, NM, August 15–17, 1999.
14. Z. Wu and B. Sundén, Heat Transfer Correlations for Elongated Bubbly Flow in Flow Boiling Micro/Minichannels, *Heat Transfer Engineering*, vol. 37, pp. 985–993, 2016.
15. W. Li and Z. Wu, A General Criterion for Evaporative Heat Transfer in Micro/Mini-Channels, *International Journal of Heat and Mass Transfer*, vol. 53, pp. 1967–1976, 2010.
16. Z. Wu and W. Li, A New Predictive Tool for Saturated Critical Heat Flux in Micro/Mini-Channels: Effect of the Heated Length-to-diameter Ratio, *International Journal of Heat and Mass Transfer*, vol. 54, pp. 2880–2889, 2011.
17. W. Li and Z. Wu, Generalized Adiabatic Pressure Drop Correlations in Evaporative Micro/Mini-Channels, *Experimental Thermal and Fluid Science*, vol. 35, pp. 866–872, 2011.
18. P.A. Kew and K. Cornwell, Correlations for the Prediction of Boiling Heat Transfer in Small Diameter Channels, *Applied Thermal Engineering*, vol. 17, pp. 705–715, 1997.
19. B. Sundén, Computational Fluid Dynamics in Research and Design of Heat Exchangers, *Heat Transfer Engineering*, vol. 28, pp. 898–910, 2007.
20. Y. Asako, T. Pi, S. Turner, and M. Faghri, Effect of Compressibility on Gaseous Flows in Micro-Channels, *International Journal of Heat and Mass Transfer*, vol. 46, pp. 3041–3050, 2003.
21. H. Sun and M. Faghri, Effects of Rarefaction and Compressibility of Gaseous flow in Micro-channel Using DSMC, *Numerical Heat Transfer, Part A*, vol. 38, pp.153–168, 2000.
22. H. Sun and M. Faghri, Effect of Surface Roughness on Nitrogen Flow in Micro-Channel Using Direct Simulation Monte Carlo, *Numerical Heat Transfer, Part A*, vol. 43, pp. 1–8, 2003.
23. M. Faghri and B. Sunden, *Heat and Fluid Flow in Microscale and Nanoscasle*, WIT Press, Southampton, UK, 2008.
24. T. Yamada, A. Kumar, Y. Asako, and M Faghri, Forced Convection Heat Transfer Simulation Using Dissipative Particle Dynamics, *Numerical Heat Transfer, Part A*, vol. 60, pp. 651–655, 2011.
25. T. Yamada, S. Hamian, B. Sundén, K. Park, and M. Faghri, Diffusive-Ballistic Heat Transport in Thin Films Using Energy Conserving Dissipative Particle Dynamics, *International Journal of Heat and Mass Transfer*, vol. 61, pp. 287–292, 2013.
26. R.E. Rudd and J.Q. Broughton, Coarse-Grained Molecular Dynamics and the Atomic Limit of Finite Elements, *Physical Review B*, vol. 58, R5893, 1998.
27. H.K. Versteeg and W. Malalasekera, *An Introduction to Computational Fluid Dynamics: The Finite Volume Method*, 2nd ed., Pearson Education, Harlow, 2007.
28. C. Hirsch, *Numerical Computation of Internal and External Flows: The Fundamentals of Computational Fluid Dynamics*, 2nd ed., Butterworth-Heinemann, Oxford, 2007.
29. M.S. Lee, V. Aute, A. Riaz, and R. Radermacher, A Review on Direct Two-Phase, Phase Change Flow Simulation Methods and Their Applications, *International Refrigeration and Air Conditioning Conference*, Purdue, July 16–19, 2012.
30. M. Wörner, Numerical Modeling of Multiphase Flows in Microfluidics and Micro Process Engineering: A Review of Methods and Applications, *Microfluidics and Nanofluidics*, vol. 12, pp. 841–886, 2012.
31. S. Patankar, *Numerical Heat Transfer and Fluid Flow*, CRC Press, Boca Raton, FL, 1980.

32. G.D. Smith, *Numerical Solution of Partial Differential Equations*, Oxford University Press, London, 1978.
33. J.N. Reddy and D.K. Gartling, *The Finite Element Method in Heat Transfer and Fluid Dynamics*, CRC Press, Boca Raton, FL, 1994.
34. H. Power and L. Wrobel, *Boundary Integral Methods in Fluid Mechanics*, Computational Mechanics Publications, Southampton, 1995.
35. G. Xia, Y. Zhai, and Z. Cui, Numerical Investigation of Thermal Enhancement in a Micro Heat Sink with Fan-Shaped Reentrant Cavities and Internal Ribs, *Applied Thermal Engineering*, vol. 58, pp. 52–60, 2013.
36. H.P. Kavepour, M. Faghri, and Y. Asako, Effects of Compressibility and Rarefaction on Gaseous Flows in Micro Channels, *Numerical Heat Transfer, Part A*, vol. 32, pp. 677–696, 1997.
37. C. Hong, Y. Asako, S. Turner, and M. Faghri, Friction Factor Correlations for Gas Flow in Slip Regime, *ASME Journal of Fluids Engineering*, vol. 129, pp. 1268–1276, 2007.
38. T. Yamada, H. Chen, M. Faghri, and O.J. Gregory, Experimental Investigations of Liquid Flow in Rib-patterned Microchannels with Different Surface Wettability, *Microfluidics and Nanofluidics*, vol. 2, pp. 83–84, 2011.
39. P. Rosa, T.G. Karayiannis, and M.W. Collins, Single-Phase Heat Transfer in Microchannels: The Importance of Scaling Effects, *Applied Thermal Engineering*, vol. 29, pp. 3447–3468, 2009.
40. G. Maranzana, I. Perry, and D. Maillet, Mini-and Micro-channels: Influence of Axial Conduction in the Walls, *International Journal of Heat and Mass Transfer*, vol. 47, pp. 3993–4004, 2004.
41. J. Li, G.P. Peterson, and P. Cheng, Three-Dimensional Analysis of Heat Transfer in a Micro-Heat Sink with Single Phase Flow, *International Journal of Heat and Mass Transfer*, vol. 47, pp. 4215–4231, 2004.
42. Y. Sui, C.J. Teo, P.S. Lee, Y.T. Chew, and C. Shu, Fluid Flow and Heat Transfer in Wavy Microchannels, *International Journal of Heat and Mass Transfer*, vol. 53, pp. 2760–2772, 2010.
43. I. Tiselj, G. Hetsroni, B. Mavko, A. Mosyak, E. Pogrebnyak, and Z. Segal, Effect of Axial Conduction on the Heat Transfer in Micro-channels, *International Journal of Heat and Mass Transfer*, vol. 47, pp. 2551–2565, 2004.
44. J. Xu, Y. Song, W. Zhang, H. Zhang, and Y. Gan, Numerical Simulations of Interrupted and Conventional Microchannel Heat Sinks, *International Journal of Heat and Mass Transfer*, vol. 51, pp. 5906–5917, 2008.
45. T. Dang, J.T. Teng, and J.C. Chu, A Study on the Simulation and Experiment of a Microchannel Counter-flow Heat Exchanger, *Applied Thermal Engineering*, vol. 30, pp. 2163–2172, 2010.
46. M. Pieper and P. Klein, A Simple and Accurate Numerical Network Flow Model for Bionic Micro Heat Exchangers, *Heat and Mass Transfer*, vol. 47, pp. 491–503, 2011.
47. J. Miwa, Y. Asako, C. Hong, and M. Faghri, Performance of Gas-to-Gas Micro-Heat Exchangers, *ASME Journal of Heat Transfer*, vol. 131, 051801, 2009.
48. S.F. Choquette, M. Faghri, M. Charmchi, and Y. Asako, Optimum Design of Microchannel Heat Sinks, *International Mechanical Engineering Congress and Exposition*, Atlanta, GA, November 17–22, 1996.
49. K. Foli, T. Okabe, M. Olhofer, Y. Jin, and B. Sendhoff, Optimization of Micro Heat Exchanger: CFD, Analytical Approach and Multi-Objective Evolutionary Algorithms, *International Journal of Heat and Mass Transfer*, vol. 49, pp. 1090–1099, 2006.
50. G. Xie, Y. Liu, B. Sundén, and W. Zhang, Computational Study and Optimization of Laminar Heat Transfer and Pressure Loss of Double-Layer Microchannels for Chip Liquid Cooling, *ASME Journal of Thermal Science and Engineering Applications*, vol. 5, 011004, 2013.

51. G. Xie, J. Liu, Y. Liu, B. Sundén, and W. Zhang, Comparative Study of Thermal Performance of Longitudinal and Transversal-Wavy Microchannel Heat Sinks for Electronic Cooling, *ASME Journal of Electronic Packaging*, vol. 135, 021008, 2013.
52. D. Attinger, C. Frankiewice, A.R. Betz, T.M. Schutzius, R. Ganguly, A. Das, C.J. Kim, and C.M. Megaridis, Surface Engineering for Phase Change Heat Transfer: A Review, *MRS Energy and Sustainability, A Review Journal*, vol. 1, E4, 2014.
53. Z. Wu, B. Sundén, W. Li, and V.V. Wadekar, Evaporative Annular Flow in Micro/Minichannels: A Simple Heat Transfer Model, *ASME Journal of Thermal Science and Engineering Applications*, vol. 5, 031009, 2013.
54. M. Wörner, *A Compact Introduction to the Numerical Modeling of Multiphase Flows*, Forschungszentrum Karlsruhe, Karlsruhe, Germany, 2003.
55. S.G. Kandlikar, Scale Effects on Flow Boiling Heat Transfer in Microchannels: A Fundamental Perspective, *International Journal of Thermal Sciences*, vol. 49, pp. 1073–1085, 2010.
56. C.E. Brennen, *Fundamentals of Multiphase Flow*, Cambridge University Press, Cambridge, 2005.
57. B.G.M. Van Wachem and A.E. Almstedt, Methods for Multiphase Computational Fluid Dynamics, *Chemical Engineering Journal*, vol. 96, pp. 81–98, 2003.
58. S.W. Welch and J. Wilson, A Volume of Fluid Based Method for Fluid Flows with Phase Change, *Journal of Computational Physics*, vol. 160, pp. 662–682, 2000.
59. M. Magnini, B. Pulvirenti, and J.R. Thome, Numerical Investigation of Hydrodynamics and Heat Transfer of Elongated Bubbles during Flow Boiling in a Microchannel, *International Journal of Heat and Mass Transfer*, vol. 59, pp. 451–471, 2013.
60. J.U. Brackbill, D.B. Kothe, and C. Zemach, A Continuum Method for Modeling Surface Tension, *Journal of Computational Physics*, vol. 100, pp. 335–354, 1992.
61. R.W. Schrage, *A Theoretical Study of Interphase Mass Transfer*, Columbia University Press, New York, 1953.
62. I. Tanasawa, Advances in Condensation Heat Transfer, *Advances in Heat Transfer*, vol. 21, pp. 55–139, 1991.
63. V.P. Carey, *Liquid-Vapor Phase-Change Phenomena*, 2nd ed., CRC Press, Boca Raton, 2007.
64. K. Okamori, *Free surface flow analysis*. Available at: http://www.cradle-cfd.com/tec/column02/002.html, accessed on February 20, 2016.
65. Z. Liu, B. Sunden, and J. Yuan, VOF Modeling and Analysis of Filmwise Condensation between Vertical Parallel Plates, *Heat Transfer Research*, vol. 43, pp. 47–68, 2012.
66. Z. Liu, B. Sunden, and H. Wu, Numerical Modeling of Multiple Bubbles Condensation in Subcooled Flow Boiling, *ASME Journal of Thermal Science and Engineering Applications*, vol. 7, 031003, 2015.
67. C.W. Hirt and B.D. Nichols, Volume of Fluid (VOF) Method for the Dynamics of Free Boundaries, *Journal of Computational Physics*, vol. 39, pp. 201–225, 1981.
68. D.L. Youngs, Time-Dependent Multi-Material Flow with Large Fluid Distortion, in *Numerical Methods for Fluid Dynamics*, ed. K.W. Morton and M.J. Baines, Academic Press, New York, 273–285, 1982.
69. Y. Renardy and M. Renardy, PROST: A Parabolic Reconstruction of Surface Tension for the Volume-of-Fluid Method, *Journal of Computational Physics*, vol. 183, pp. 400–421, 2002.
70. Y.Q. Zu, Y.Y. Yan, S. Gedupudi, T.G. Karayiannis, and D.B.R. Kenning, Confined Bubble Growth during Flow Boiling in a Mini-/Micro-Channel of Rectangular Cross-Section Part II: Approximate 3-D Numerical Simulation, *International Journal of Thermal Sciences*, vol. 50, pp. 267–273, 2011.
71. R. Zhuan and W. Wang, Flow Pattern of Boiling in Micro-Channel by Numerical Simulation, *International Journal of Heat and Mass Transfer*, vol. 55, pp. 1741–1753, 2012.

72. H. Ganapathy, A. Shooshtari, K. Choo, S. Dessiatoun, M. Alshehhi, and M. Ohadi, Volume of Fluid-based Numerical Modeling of Condensation Heat Transfer and Fluid Flow Characteristics in Microchannels, *International Journal of Heat and Mass Transfer*, vol. 65, pp. 62–72, 2013.

73. E. Da Riva and D. Del Col, Numerical Simulation of Laminar Liquid Film Condensation in a Horizontal Circular Minichannel, *ASME Journal of Heat Transfer*, vol. 134, 051019, 2012.

74. Z. Yin, Y. Guo, B. Sunden, Q. Wang, and M. Zeng, Numerical Simulation of Laminar Film Condensation in a Horizontal Minitube with and Without Non-Condensable Gas by the VOF Method, *Numerical Heat Transfer, Part A: Applications*, vol. 68, pp. 958–977, 2015.

75. M. Sussman, P. Smereka, and S. Osher, A Level Set Approach for Computing Solutions to Incompressible Two-phase Flow, *Journal of Computational Physics*, vol. 114, pp. 146–159, 1994.

76. G. Son, V.K. Dhir, and N. Ramanujapu, Dynamics and Heat Transfer Associated with a Single Bubble during Nucleate Boiling on a Horizontal Surface, *ASME Journal of Heat Transfer*, vol. 121, pp. 623–631, 1999.

77. A. Mukherjee and V.K. Dhir, Study of Lateral Merger of Vapor Bubbles during Nucleate Pool Boiling, *ASME Journal of Heat Transfer*, vol. 126, pp. 1023–1039, 2004.

78. A. Mukherjee and S.G. Kandlikar, Numerical Simulation of Growth of a Vapor Bubble during Flow Boiling of Water in a Microchannel, *Microfluidics and Nanofluidics*, vol. 1, pp. 137–145, 2005.

79. A. Mukherjee, S.G. Kandlikar, and Z.J. Edel, Numerical Study of Bubble Growth and Wall Heat Transfer during Flow Boiling in a Microchannel, *International Journal of Heat and Mass Transfer*, vol. 54, pp. 3702–3718, 2011.

80. Y. Suh, W. Lee, and G. Son, Bubble Dynamics, Flow, and Heat Transfer during Flow Boiling in Parallel Microchannels, *Numerical Heat Transfer, Part A: Applications*, vol. 54, pp. 390–405, 2008.

81. M. Sussman and E.G. Puckett, A Coupled Level Set and Volume-of-Fluid Method for Computing 3D and Axisymmetric Incompressible Two-Phase Flows, *Journal of Computational Physics*, vol. 162, pp. 301–337, 2000.

82. Q. Liu and B. Palm, A Numerical Study of Bubble Growing during Saturated and Sub-Cooled Flow Boiling in Micro Channels, *4th Micro and Nano Flows Conference*, University College London, September 7–10, 2014.

83. B.A. Nichita, *An Improved CFD Tool to Simulate Adiabatic and Diabatic Two-Phase Flows*, PhD thesis, EPFL, Lausanne, Switzerland, 2010.

84. M. Herrmann, *Refined Level Set Grid Method for Tracking Interfaces. Annual Research Briefs, Center for Turbulence Research*, Stanford University, Stanford, CA, 2005.

85. S.O. Unverdi and G. Tryggvason, A Front-Tracking Method for Viscous, Incompressible, Multi-Fluid Flows, *Journal of Computational Physics*, vol. 100, pp. 25–37, 1992.

86. D. Juric and G. Tryggvason, Computations of Boiling Flows, *International Journal of Multiphase Flow*, vol. 24, pp. 387–410, 1998.

87. G. Tryggvason, B. Bunner, A. Esmaeeli, D. Juric, N. Al-Rawahi, W. Tauber, J. Han, S. Nas, and Y.J. Jan, A Front-tracking Method for the Computations of Multiphase Flow, *Journal of Computational Physics*, vol. 169, pp. 708–759, 2001.

88. T. Ye, W. Shyy, and J.N. Chung, A Fixed-Grid, Sharp-Interface Method for Bubble Dynamics and Phase Change, *Journal of Computational Physics*, vol. 174, pp. 781–815, 2001.

89. J.P. Gois, A. Nakano, L.G. Nonato, and G.C. Buscaglia, Front Tracking with Moving-Least-Squares Surfaces, *Journal of Computational Physics*, vol. 227, pp. 9643–9669, 2008.

90. X. He and G.D. Doolen, Thermodynamic Foundations of Kinetic Theory and Lattice Boltzmann Models for Multiphase Flows, *Journal of Statistical Physics*, vol. 107, pp. 309–328, 2002.

91. T. Inamuro, T. Ogata, S. Tajima, and N. Konishi, A Lattice Boltzmann Method for Incompressible Two-Phase Flows with Large Density Differences, *Journal of Computational Physics*, vol. 198, pp. 628–644, 2004.

92. H.W. Zheng, C. Shu, Y.T. Chew, and J.H. Sun, Three-Dimensional Lattice Boltzmann Interface Capturing Method for Incompressible Flows, *International Journal for Numerical Methods in Fluids*, vol. 56, pp. 1653–1672, 2008.

93. S. Gong, P. Cheng, and X. Quan, Lattice Boltzmann Simulation of Droplet Formation in Microchannels under an Electric Field, *International Journal of Heat and Mass Transfer*, vol. 53, pp. 5863–5870, 2010.

94. S. Gong and P. Cheng, A Lattice Boltzmann Method for Simulation of Liquid–Vapor Phase-Change Heat Transfer, *International Journal of Heat and Mass Transfer*, vol. 55, pp. 4923–4927, 2012.

95. S. Gong and P. Cheng, Numerical Investigation of Saturated Flow Boiling in Microchannels by the Lattice Boltzmann Method, *Numerical Heat Transfer, Part A: Applications*, vol. 65, pp. 644–661, 2014.

96. V. Talimi, Y.S. Muzychka, and S. Kocabiyik, A Review on Numerical Studies of Slug Flow Hydrodynamics and Heat Transfer in Microtubes and Microchannels, *International Journal of Multiphase Flow*, vol. 39, pp. 88–104, 2012.

97. C. Chung, K.H. Ahn, and S.J. Lee, Numerical Study on the Dynamics of Droplet Passing through a Cylinder Obstruction in Confined Microchannel Flow, *Journal of Non-Newtonian Fluid Mechanics*, vol. 162, pp. 38–44, 2009.

98. C. Fang, M. David, A. Rogacs, and K. Goodson, Volume of Fluid Simulation of Boiling Two-Phase Flow in a Vapor-Venting Microchannel, *Frontiers in Heat and Mass Transfer*, vol. 1, 013002, 2010.

99. S. Nebuloni and J.R. Thome, Numerical Modeling of Laminar Annular Film Condensation for Different Channel Shapes, *International Journal of Heat and Mass Transfer*, vol. 53, pp. 2615–2627, 2010.

100. W. Wang, Z. Liu, Y. Jin, and Y. Cheng, LBM Simulation of Droplet Formation in Micro-Channels, *Chemical Engineering Journal*, vol. 173, pp. 828–836, 2011.

101. W. Lee, G. Son, and H.Y. Yoon, Direct Numerical Simulation of Flow Boiling in a Finned Microchannel, *International Communications in Heat and Mass Transfer*, vol. 39, pp. 1460–1466, 2012.

102. S. Nebuloni and J.R. Thome, Numerical Modeling of the Conjugate Heat Transfer Problem for Annular Laminar Film Condensation in Microchannels, *ASME Journal of Heat Transfer*, vol. 134, 051021, 2012.

103. M. Magnini, B. Pulvirenti, and J.R. Thome, Numerical Investigation of the Influence of Leading and Sequential Bubbles on Slug Flow Boiling within a Microchannel, *International Journal of Thermal Sciences*, vol. 71, pp. 36–52, 2013.

104. M.J. Miner and P.E. Phelan, Effect of Cross-Sectional Perturbation on Critical Heat Flux Criteria in Microchannels, *ASME Journal of Heat Transfer*, vol. 135, 101009, 2013.

105. S. Chen, Z. Yang, Y. Duan, Y. Chen, and D. Wu, Simulation of Condensation Flow in a Rectangular Microchannel, *Chemical Engineering and Processing: Process Intensification*, vol. 76, pp. 60–69, 2014.

106. W. Lan, S. Li, Y. Wang, and G. Luo, CFD Simulation of Droplet Formation in Microchannels by a Modified Level Set Method, *Industrial & Engineering Chemistry Research*, vol. 53, pp. 4913–4921, 2014.

107. S.M. Kim, J. Kim, and I. Mudawar, Flow Condensation in Parallel Micro-Channels– Part 1: Experimental Results and Assessment of Pressure Drop Correlations, *International Journal of Heat and Mass Transfer*, vol. 55, pp. 971–983, 2012.

108. H.Y. Wu and P. Cheng, Condensation Flow Patterns in Silicon Microchannels, *International Journal of Heat and Mass Transfer*, vol. 48, pp. 2186–2197, 2005.

109. J.S. Hu and C.Y. Chao, An Experimental Study of the Fluid Flow and Heat Transfer Characteristics in Micro-Condensers with Slug-Bubbly Flow, *International Journal of Refrigeration*, vol. 30, pp. 1309–1318, 2007.

110. J.W. Coleman and S. Garimella, Two-phase Flow Regimes in Round, Square and Rectangular Tubes during Condensation of Refrigerant R134a, *International Journal of Refrigeration*, vol. 26, pp. 117–128, 2003.

111. S. Gong and P. Cheng, Numerical Investigation of Droplet Motion and Coalescence by an Improved Lattice Boltzmann Model for Phase Transitions and Multiphase Flows, *Computers & Fluids*, vol. 53, pp. 93–104, 2012.

5 Review of Advances in Heat Pipe Analysis and Numerical Simulation

Amir Faghri and Theodore L. Bergman

CONTENTS

ABSTRACT: There has been a major transformation in heat pipe science and technology and its application over the last several decades due to the development and advancement of new heat pipes as well as large-scale commercialization and manufacturing. Heat pipe analysis and numerical simulation have been improved using state-of-the-art computational modeling and better multiphase physical understanding due to analysis and consideration of the vast experimental data available on heat pipes under various operating conditions. A review of heat pipe analysis and numerical simulation covering all types of heat pipes with various levels of approximation is presented. These models can provide an effective design tool for application of heat pipes under different operating conditions, including integration with other thermal devices and systems.

5.1   INTRODUCTION

Heat pipes are highly effective thermal devices with attractive performance and operating characteristics (Faghri 1995, 2012, 2014; Shabgard et al. 2015). They offer considerable, near-isothermal heat transport across large distances while providing extensive design freedom, simple and inexpensive construction, and fine thermal control with no external pumping power requirement. Benefits of heat pipes over conventional heat transfer methods include the ability to conduct high heat rates through a small cross-sectional area, at various temperature levels, and with minimal end-to-end temperature drops.

Current technological advancements and new applications, including lower manufacturing costs, have led to a surge of interest in heat pipes over the last several decades as well as to the development of new types of heat pipes. Specifically, the critical need for superior cooling of electronics and energy systems has required a large-scale increase in the manufacturing and design of heat pipes. Several million heat pipes per month are now being manufactured solely for the purpose of transferring heat away from and cooling the processors of laptop computers. Furthermore, research and development of "nonconventional" heat pipes, such as loop heat pipes, micro and miniature heat pipes, and pulsating heat pipes, has progressed enough for practical application.

There is a wealth of published literature for those interested in heat pipe science and technology and the current state of the field, including several heat pipe books and monographs, as well as the proceedings of eleven international heat pipe symposiums and seventeen international heat pipe conferences. Additional archival and nonarchival literature is widely available that spans the last three decades. Fundamental and applied research and development has been performed through numerous efforts over the last three decades and remains a primary focus today because of the unique potential benefits afforded by heat pipes.

5.2   PRINCIPLES OF OPERATION

The basic components and operational principle of a conventional heat pipe are illustrated in Figure 5.1a using a cylindrical geometry, though they may be of any size or shape. A heat pipe is usually comprised of a sealed container including the pipe wall and end caps, a wick structure, and a small amount of working fluid at saturation pressure (the pressure at which the vapor is in thermodynamic equilibrium with its condensed phase). Commonly used working fluids include water, acetone, methanol, ammonia, and sodium, and they may be chosen based on the operating temperature required by the system (Faghri 2014). The length of the heat pipe may also be divided into three regions with characteristic behavior and boundary conditions: the evaporator section, adiabatic (transport) section, and the condenser section. Modifications may also include multiple heat sources or sinks, and the adiabatic section may not be required, depending on the application and overall design.

An external heat source is applied to the evaporator section, and the energy is transferred to the wick structure by a combination of conduction and convection. The working fluid in the wick is then vaporized, and the resulting vapor pressure drives the vapor through the adiabatic section and to the condenser. The vapor condenses and releases

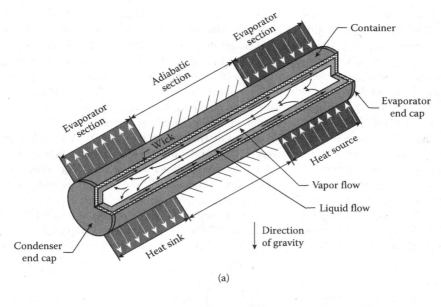

(a)

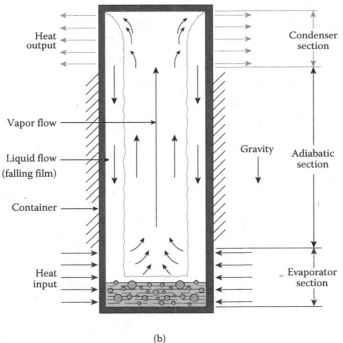

(b)

FIGURE 5.1 Schematic of a conventional (a) heat pipe and (b) thermosyphon showing the principle of operation and circulation of the working fluid.

its latent heat of vaporization to an external heat sink located adjacent to the condenser section, ultimately rejecting heat energy from the original heat input with minimal loss. Simultaneously, the condensed liquid develops a capillary pressure due to the menisci in the wick, which serves to pump the liquid back to the evaporator section. The heat pipe can, therefore, transport the latent heat of vaporization from the evaporator to the condenser section and will continue to do so as long as there is sufficient capillary pressure to replenish the evaporator with liquid. A conventional two-phase closed thermosyphon (Figure 5.1b) is similar in construction and operation to a heat pipe, though it lacks a wick structure; instead, a thermosyphon uses gravity-assisted falling films of condensate and thus requires the evaporator to be below the condenser section. Thermosyphons are generally used in many applications with no wick requirement.

The axial vapor pressure in a heat pipe is influenced by frictional, inertial, blowing (evaporation), and suction (condensation) effects, while the axial change in the liquid pressure is mainly due to friction. Typical axial variation of the internal liquid and vapor pressures is presented in Figure 5.2 for low, moderate, and high vapor flow rates (Figure 5.2a–c, respectively). However, when additional body forces are present on

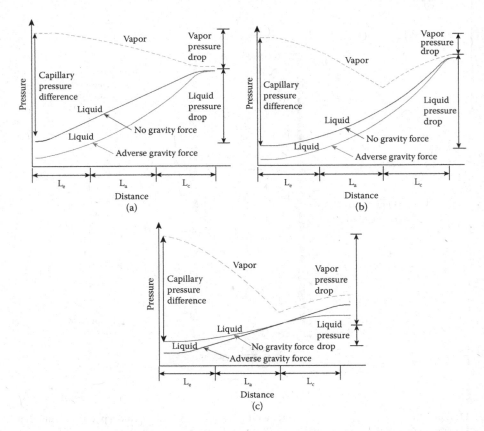

FIGURE 5.2 Axial variation of the liquid–vapor interface and the vapor and liquid pressures along the heat pipe at (a) low, (b) moderate, and (c) high vapor flow rates.

the heat pipe, such as an adverse gravitational force, the liquid pressure drop is larger. Therefore, the capillary pressure must be higher as well, in order to sufficiently return the liquid to the evaporator at a given heat input. For moderate vapor flow rates, the axial vapor drop and recovery along the condenser section are due to dynamic effects (Figure 5.2b). For a heat pipe with a high vapor flow rate, the vapor and liquid pressure drops in the condenser section are different as compared to heat pipes of low and moderate vapor flow rates (Figure 5.2c). The vapor pressure drop can surpass the liquid pressure drop; therefore, the liquid pressure may be higher than the vapor pressure in the condenser section if the liquid and vapor pressures are approximately equal at the end cap.

Traditional analysis of heat pipes relies on fundamental fluid mechanics and heat transfer principles. The axial liquid pressure drop in the wick structure, the maximum capillary pumping power, and the vapor flow in the vapor core may all be described with existing concepts in fluid mechanics. Similarly, both the internal and external thermal behaviors can be described with heat transfer and multiphase theory. Conjugate heat conduction in the wall and the wick, evaporation and condensation at the liquid–vapor interface, and forced convection in the vapor channel and wick may all be accurately described. A thermal network approach may be adopted for simple heat pipe analysis, and the thermal resistances (or elements, for transient analysis) in a conventional heat pipe and thermosyphon are presented in Figure 5.3. Other associated thermal processes may be important for transient modeling of a heat pipe, such as solidification and liquefaction, and rarefied gas theory may be required for frozen start-up analysis.

The freedom in design for heat pipes is made apparent by the considerable dimensional variability for all heat pipe types and shapes. Micro heat pipes with dimensions as low as 30 μm wide by 80 μm deep by 19.75 mm long, and large-scale heat pipes with lengths of 100 m have been constructed (Faghri 2014). All heat pipes have the basic requirement of at least an evaporator and condenser section, where the working fluid evaporates and condenses, respectively. The additional adiabatic (or transport) section separates the evaporator and condenser sections by an appropriate distance and may be included to satisfy heat pipe limitations or the design constraints of a particular application. Heat pipes may have multiple evaporators, condensers, or adiabatic sections, and the working fluid is usually circulated by capillary forces in the wick structure. Gravitational, centrifugal, and osmotic forces are also used to aid in condensate return to the evaporator section (Faghri 2012).

Traditionally, conventional heat pipes have been constructed as circular cylinders. However, geometries that are more complex to design and manufacture have been studied and constructed, such as rectangular (flat heat pipes), conical (rotating heat pipes), corrugated flexible heat pipes, and nosecap (leading edge) heat pipes (Faghri 2014).

5.3 HEAT PIPE ANALYSIS AND NUMERICAL SIMULATION

The internal flow mechanics, heat transfer rate, and performance characteristics of a heat pipe may be determined from the liquid and vapor pressure drops in the separate segments. The fundamental mechanisms of heat transfer in a heat pipe can be divided into four categories: (1) vapor flow in the vapor channel, (2) liquid flow in the wick, (3) interaction between the liquid and vapor flows, and (4) heat conduction in the wall. Most of the analytical and numerical efforts have been focused on the

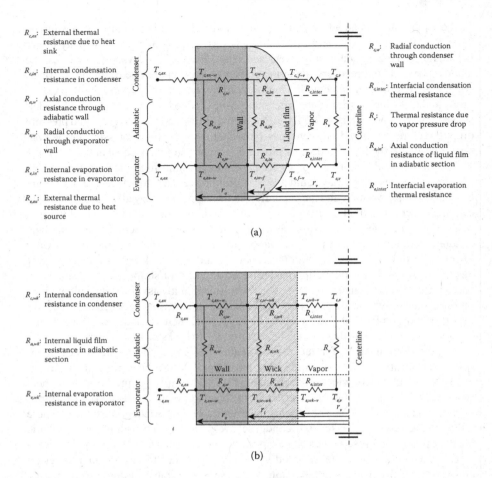

FIGURE 5.3 Thermal resistance model of a conventional (a) thermosyphon and (b) heat pipe. (Reprinted from *International Journal of Heat and Mass Transfer*, 89, Shabgard, H., Allen, M. J., Sharifi, N., Benn, S. P., Faghri, A. and Bergman, T. L., Heat Pipe Heat Exchangers and Heat Sinks: Opportunities, Challenges, Applications, Analysis, and State of the Art, 138–158, Copyright (2015), with permission from Elsevier.)

vapor core region and heat conduction in the wall of the heat pipe due to the inherent difficulty and complexity of most theoretical models for simulation of the liquid flow. The presence of the wick structure complicates analysis of the liquid flow and the liquid–vapor interaction and generally requires empirical information that is only obtained from experimentation. Further complexities of analyses are generally due to the various shapes made possible for heat pipes. While the liquid flow in the wick and liquid–vapor interactions (categories 2 and 3, respectively) are basically similar despite the overall shape of the heat pipe, the dynamics of the vapor flow and the heat conduction in the wall (categories 1 and 4, respectively) are dependent on the geometry and boundary conditions for nonconventional heat pipes.

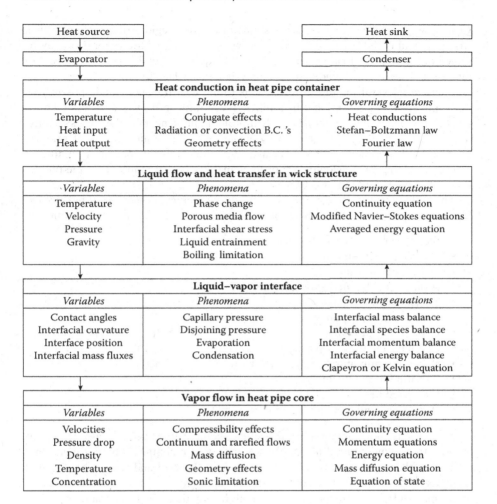

FIGURE 5.4 Chart diagram for heat pipe simulation and interaction among different regions.

The physical phenomena and interactions among the regions of a conventional heat pipe are best illustrated by Figure 5.4. The chart diagram indicates that only physically adjacent components of the heat pipe may interact with one another. For example, the container/wall may not interact with the vapor region directly but may do so via the wick structure and the liquid–vapor interface. In general, the heat pipe regions represented in the chart diagram should all be solved as a conjugate problem. However, for real application, approximations are often introduced to simplify the complex nature of the analysis. These approximations may include neglecting or combining specific regions or decoupling the regions from one another. It may be suitable to neglect the liquid flow in the wick, for example, in which the wick and heat pipe container regions may be considered together, with the mechanism of heat transfer in both regions being mainly pure conduction. Alternatively, if only the vapor flow of the heat pipe is of interest, the three other regions (the container, wick structure, and liquid–vapor interface) may be neglected, and the vapor flow may be solved independently. In some applications, the liquid flow in the wick structure is of interest and should be included in the analysis for capillary limit calculations.

There are several important restrictions to the heat transport made possible by the vapor flow in conventional heat pipes, including capillary, sonic, and entrainment limits. The sonic limitation is only minimally influenced by the wick structure; therefore, the choking phenomenon associated with the sonic limit is of primary interest for vapor flow analysis because the vapor velocity becomes significant compared to the sonic velocity. Vapor flow analysis is also required for the prediction of the capillary limitation of a heat pipe.

The following sections summarize the developments in modeling and discuss important results for different types of heat pipes under various operating conditions:

- Steady-state heat pipe analysis
- Transient heat pipe analysis
- Frozen start-up analysis
- Axially grooved heat pipe analysis
- Thermosyphon analysis
- Rotating heat pipe analysis
- Loop heat pipe analysis
- Capillary pumped loop heat pipe analysis
- Micro and miniature heat pipe analysis
- Pulsating heat pipe analysis
- Integrated heat pipes and phase change materials
- Heat pipes with nanofluids and nanoparticles

For analysis and simulation for each situation, there are varying levels of approximation, from simple one-dimensional vapor flow to a complete three-dimensional analysis with inclusion of the conjugate nature of the wall and wick. The different methods for heat pipe analysis utilize different approximations and approaches, such as compressible versus incompressible, analytical and closed-form solutions versus numerical analyses, as well as one-, two-, and three-dimensional modeling of each region. Other important modeling considerations include heat pipe geometry, heat input distributions, and various other boundary conditions.

5.4 STEADY-STATE HEAT PIPE ANALYSIS

An accurate prediction of operational performance at steady-state conditions is of significant value in the design of heat pipes. Faghri (1986) modeled the steady-state two-dimensional incompressible vapor flow in annular and conventional heat pipes. In the study, the elliptic conservation equations were reduced to partially parabolic Navier–Stokes equations using boundary layer approximations, which were solved using a fully implicit, marching finite-difference scheme.

This work was later extended by Faghri and Parvani (1988) and Faghri (1989) with the inclusion of vapor compressibility and solution of fully elliptic Navier–Stokes equations. One- and two-dimensional models were developed for solution of the vapor flow in annular and conventional heat pipes. They also provided analytical expressions for the axial pressure drop within both annular and conventional heat pipes.

Block-heated heat pipes are often applied for efficient heat transfer and energy conservation in various thermal systems and in electronic cooling. These systems are defined by the relatively nonuniform heat input applied by the block heater at the evaporator section of each heat pipe. Cao et al. (1989) provided a simplified model for block-heated heat pipes by dividing the evaporator section into two separately classified regions, one with the applied heat flux from the block heater and another section describing the unheated portion of the wall. The simulation results were compared to the heat pipe studied experimentally by Rosenfeld (1987).

The appreciable effect of conjugate heat transfer within the heat pipe wall and wick was demonstrated by Chen and Faghri (1990). Coupled, two-dimensional domains of the compressible vapor flow and conduction of the wall and wick were modeled, and the complete elliptic Navier–Stokes equations were solved for the vapor flow using a finite control volume iterative technique. The simulation was validated using the experimental data provided by Ivanovskii et al. (1982) for a high-temperature sodium/stainless-steel heat pipe (Figure 5.5). Three different methods were presented, where the vapor flow was described as compressible elliptic, incompressible elliptic, and compressible parabolic. The compressible elliptic and compressible parabolic solutions show reasonable vapor temperature distributions as compared to the experimental data.

Faghri and Buchko (1991) extended a similar methodology to model a low-temperature, heat pipe with multiple heat sources. A two-dimensional compressible vapor flow was coupled to a two-dimensional conduction model in the wall.

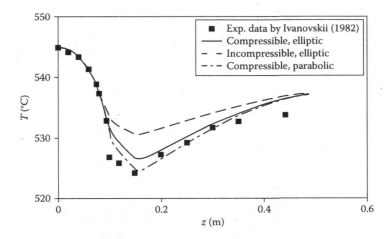

FIGURE 5.5 The axial interface temperature profile along the sodium heat pipe with $Q = 560$ W, $R_v = 0.007$ m, $L_e = 0.1$ m, $L_a = 0.05$ m, $k_l = 66.2$ W/m^2-K, $k_s = 19.0$ W/m^2-K, $\delta_l = 0.0005$ m, and $\delta_w = 0.001$ m. (Reprinted from *International Journal of Heat and Mass Transfer*, 33, Chen, M. M. and Faghri, A., An Analysis of the Vapor Flow and the Heat Conduction through the Liquid-Wick and Pipe Wall in a Heat Pipe with Single Or Multiple Heat Sources, 1945–1955, Copyright (1990), with permission from Elsevier.)

The model was made more comprehensive, however, by including the two-dimensional effects of liquid flow in the wick using volume-averaged velocities in the porous media, and the conservation equations were solved using an elliptic finite control volume iterative scheme. Some of the results are presented for two different situations in Figure 5.6. In the first case, only Evaporator 1 was active, with a heat input of 97 W. In the second case, Evaporators 1 and 2 were active, with heat inputs of 99 W and 98 W, respectively. The results of the study were compared to the experimental data from a water–copper heat pipe with multiple heat sources.

The thermal behavior of a flat plate heat pipe was examined analytically by Zhu and Vafai (1998b). The vapor and liquid flows were considered for steady and

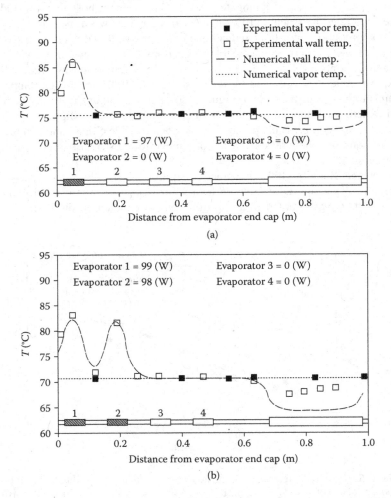

FIGURE 5.6 Heat pipe wall and vapor temperature versus axial location for (a) single evaporator and (b) two evaporators. (Reprinted from *Journal of Heat Transfer*, Faghri, A. & Buchko, M., Experimental and Numerical Analysis of Low-Temperature Heat Pipes with Multiple Heat Sources, 1991, with permission from ASME.)

incompressible conditions, and the pseudo-three-dimensional analytical model used boundary layer approximation to describe the vapor flow. The steady-state operation of a flat plate miniature heat pipe with multiple evaporators was simulated by Lefèvre and Lallemand (2006). They used analytical solutions for the two-dimensional hydrodynamic description of the liquid and vapor flows with a three-dimensional heat conduction model in the wall. Furthermore, the model was able to predict the maximum heat transfer capability for the miniature heat pipe.

Xiao and Faghri (2008) analyzed the thermal and hydrodynamic behavior of flat heat pipes and vapor chambers without empirical correlations. The detailed, three-dimensional model accounted for heat conduction in the wall, fluid flow in the vapor chamber and porous wicks, and the coupled heat and mass transfer at the liquid–vapor interface. The numerical results were compared with the experimental data from Wang and Vafai (2000), as shown in Figure 5.7, where the temperature distributions at the surface of the top and bottom walls are presented. The deviation in the temperatures of the condenser section at the top wall is mainly due to local variation in the convective heat transfer coefficient due to the variable surface temperatures in the experiment, while the surface heat transfer coefficient was assumed constant in the numerical model.

Aghvami and Faghri (2011) studied various heating and cooling configurations of flat plate heat pipes and vapor chambers using a steady-state analytical thermal-fluid model that included the wall, liquid, and vapor domains. The parametric investigation indicated that uniform evaporation and condensation in the axial direction may only be accurate if the wall thermal conductivity is small. Shabgard and Faghri (2011) extended the model for cylindrical heat pipes with multiple evaporators. Two-dimensional heat conduction in the heat pipe wall was coupled to the liquid flow in the wick and the vapor flow mechanics. Results showed that neglecting the axial heat conduction through the wall may cause an overestimation of pressure drops by up to 10%, depending on the heat pipe specifications.

5.5 TRANSIENT HEAT PIPE ANALYSIS

A numerical two-dimensional transient heat pipe analysis from Cao and Faghri (1990) accounted for both vapor compressibility and coupled vapor flow and heat conduction in the wall and wick. A high-temperature heat pipe was subjected to a pulsed heat input, with either convective or radiative cooling at the condenser section. It was concluded that, for a high-temperature heat pipe, the pure conduction model was sufficient to describe heat transfer in the wick. The transient vapor temperature profile for a pulsed heat input from 623 to 770 W is included in Figure 5.8a, with a convective boundary condition at the condenser section. The operating temperature of the heat pipe increases with time after the heat pulse due to the elevation in temperature of the outer pipe wall temperature and the coupled nature of the vapor flow. The outer pipe wall of the condenser section must increase with time to reject additional heat. The transient axial vapor temperature profile with the radiative boundary condition is similarly shown in Figure 5.8b. Faghri et al. (1991b) expanded on the model of Cao and Faghri (1990) to include multiple heat sources and sinks. The numerical results for continuum transient and

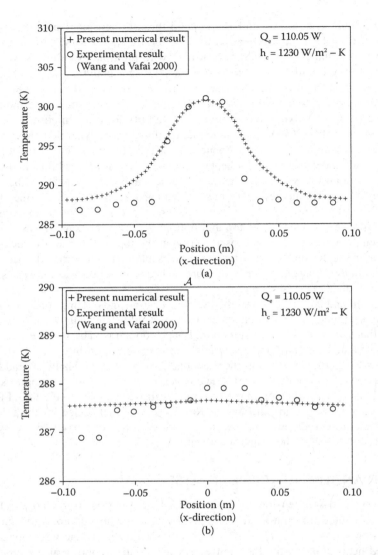

FIGURE 5.7 Verification of numerical results for temperature distribution along a flat heat pipe with vertical wick columns at (a) the top wall and (b) the bottom wall at Z = W/2. (Reprinted from *International Journal of Heat and Mass Transfer*, 51, Xiao, B. and Faghri, A., A Three-Dimensional Thermal-Fluid Analysis of Flat Heat Pipes, 3113–3126, Copyright (2008), with permission from Elsevier.)

steady-state conditions for a high-temperature heat pipe with multiple heat sources were compared with the experimental results of Faghri et al. (1991a) and found to be in good agreement.

A method for the transient analysis of nonconventional heat pipe geometries was proposed by Cao and Faghri (1991, 1993a). The one-dimensional compressible

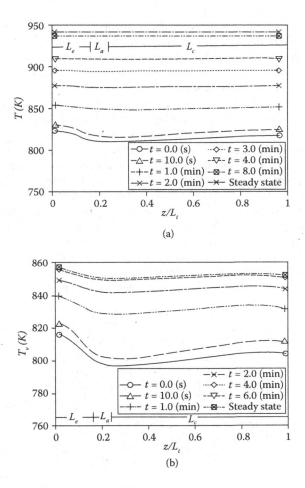

FIGURE 5.8 Centerline vapor temperature for transient response to heat input pulse: (a) convective boundary condition and (b) radiative boundary condition. (Reprinted from *Journal of Heat Transfer*, Cao, Y. and Faghri, A., A Numerical Analysis of High-Temperature Heat Pipe Startup from the Frozen State, 1990, with permission from ASME.)

vapor flow was coupled to the two- or three-dimensional heat conduction in the wall and wick. The model simplified the flow of vapor to one dimension, which required friction coefficient information for laminar and turbulent flow regimes. For the complex geometries considered in the study, such as the leading edge of a wing or a spacecraft nosecap, a multidimensional formulation is used in the wall and wick. Conduction in the wall and wick is coupled to the vapor flow via conjugate heat transfer, and the elliptic conservation equations of the vapor are solved. The liquid flow in the wick was assumed to be decoupled from the heat pipe operation in order to find the capillary limit for the nonconventional geometries.

Zuo and Faghri (1998) used a lumped model and thermal network approach for transient heat pipe analysis. The heat pipe was reduced to a number of components involving different thermal resistances and elements. The governing equations of the heat pipe were simplified to a set of first-order, linear, differential equations. Zhu and Vafai (1998a) presented an analytical solution for the quasi-steady and incompressible vapor flow with transient one-dimensional heat conduction in the wall and wick for an asymmetrical flat plate heat pipe.

Rice and Faghri (2007) formulated a detailed two-dimensional transient numerical analysis of flat and cylindrical heat pipes with screen wicks. The model included flow in the wick and was performed independently of any empirical correlations. Both single and multiple heat sources were used as heat inputs, and the condenser section was cooled with constant heat flux as well as convective and radiative heat sinks. An additional important consideration is that the internal pressure was not fixed based on any reference point but was allowed to float based on the physics of the problem. Accordingly, the capillary pressure required to drive the flow of liquid in the wick was obtained for the various heating configurations and operating conditions. The capillary pressures were used in an analysis for the prediction of the maximum capillary pressure for a given heat load and to determine the dryout limitations of a heat pipe. The wick and vapor interface requires special attention because of the heterogeneous interactions due to surface tension and mass transfer (Figure 5.9).

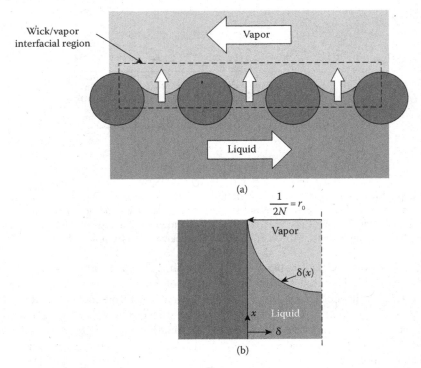

FIGURE 5.9 Schematic of (a) a wick–liquid–vapor interfacial region and (b) an idealized single pore. (From Analysis of Porous Wick Heat Pipes, Including Capillary Dry-Out Limitations, Rice, J. and Faghri A., 2007, reprinted by permission of the American Institute of Aeronautics and Astronautics.)

The transient models presented earlier may predict the performance of heat pipes at various operating conditions, and of different wick structures, by neglecting the curvature effect at the liquid–vapor interface. However, as the heat pipe size or wick thickness decreases into the micro range, the detailed heat and mass transport may become important in determining the capillary limit, especially for wicks with high thermal conductivities. Several investigators, including Tournier and El-Genk (1994), Rice and Faghri (2007), and Ranjan et al. (2011), have made efforts to include the effects of the radius of curvature of the liquid menisci formed in the screen wick pores in full numerical simulations and to predict the capillary limit in a heat pipe.

5.6 FROZEN START-UP ANALYSIS

Some applications may require heat pipes to transfer energy but initially may have a working fluid in the solid phase—either due to the working fluid properties (such as sodium) or because of the operating conditions of the system. As the heat transfer progresses, the frozen working fluid melts, and the heat pipe transitions to normal operation. As a major consideration for heat pipe systems and operation, frozen start-up has been a primary interest to be modeled via full numerical simulation. Description of the frozen start-up process includes the consideration of the working fluid as solid, mushy, or liquid, and the vapor flow as free molecular, partially continuum, continuum transient, and continuum steady. In the analysis by Jang et al. (1990), the heat transfer through the free molecular vapor flow was neglected.

Early in the start-up period, the vapor in the heat pipe vapor channel may be considered to be at a rarefied state. This occurs when the vapor has an extremely low vapor density, and it partly loses its continuum characteristics. During start-up, when the melting interface is within the wick and has not yet reached the wick–vapor interface, the liquid–vapor interface is not adiabatic. This is an important distinction because an adiabatic liquid–vapor interface would mean that there is no vapor accumulation in the vapor space and that the vapor flow has not reached the continuum state. Cao and Faghri (1993b) simulated the rarefied vapor flow using a self-diffusion model, where self-diffusion refers to the interdiffusion of particles of the same species due to the density gradient. Heat transfer through the rarefied vapor flow was coupled to the phase change of the working fluid in the wick but was considered to be valid only for early stages of frozen start-up.

Cao and Faghri (1993c) developed a complete numerical simulation of frozen start-up by combining the rarefied self-diffusion model (Cao and Faghri 1993b) with the continuum transient model of Cao and Faghri (1990). Frozen start-up was completely modeled in all regions, including the effects of conjugate heat transfer within the wall. Heat transfer and fluid flow in the wick were coupled to the vapor flow. Numerical simulation of a multiple-evaporator high-temperature sodium/stainless-steel heat pipe was performed and compared to experimental results from Faghri et al. (1991a), shown in Figure 5.10a, as well as to the sodium heat pipe studied by Ponnappan (1990) (see Figure 5.10b). Comparison to the experimental data is excellent, and the location and progression of the hot front is closely simulated as a function of time.

Frozen start-up of a high-temperature heat pipe is dependent on a number of factors, including the difference between the melting temperature of the working fluid

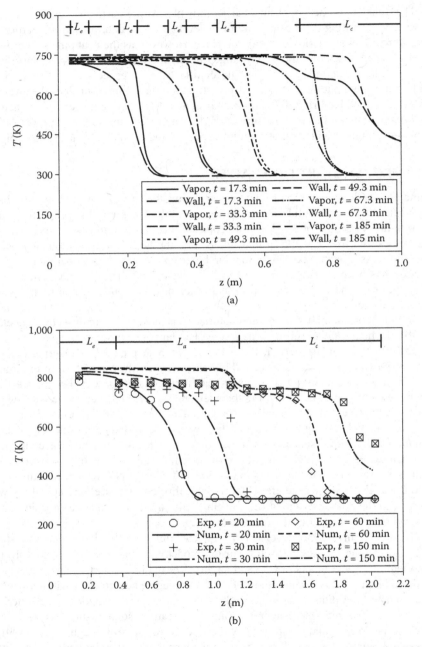

FIGURE 5.10 Wall temperature prediction for frozen start-up by Cao and Faghri (1993c) compared with the experimental data of (a) Faghri et al. (1991a) and (b) Ponnappan (1990). (Reprinted from *Journal of Heat Transfer*, Cao, Y. and Faghri, A., A Numerical Analysis of High-Temperature Heat Pipe Startup from the Frozen State, 1993c, with permission from ASME.)

and the ambient temperature, the liquid density of the working fluid, the porosity of the wick, and the physical dimensions of the heat pipe. The frozen start-up limit (FSL) is a nondimensional parameter defined by Cao and Faghri (1992) that measures the ability of a heat pipe to start-up from the frozen state, and it has been validated from various experimental cases. Determination of the frozen start-up limitation and quantification using the FSL is one of the most significant contributions to frozen start-up heat pipe analysis. The previous two-dimensional models offered accurate prediction of frozen start-up of low- and high-temperature heat pipes. However, the models assumed a flat liquid–vapor interface. Investigations including Hall et al. (1994) and Tournier and El-Genk (1996) assumed one-dimensional vapor flow regimes and accounted for the local interfacial radius of curvature or the liquid meniscus.

5.7   AXIALLY GROOVED HEAT PIPE ANALYSIS

Axially grooved heat pipes (AGHPs) involve modeling of the fluid circulation along with the heat and mass transfer associated with evaporation and condensation. Khrustalev and Faghri (1995a) mathematically modeled low-temperature AGHPs, while the model by Khrustalev and Faghri (1995b) accounted for the important contributions of heat transfer in the microfilm region. The heat transfer through the liquid film and the fin between the grooves in the evaporator and condenser section was examined by including the effects of the disjoining pressure and interfacial thermal resistance and surface roughness. The model used one-dimensional axial heat transfer in the heat pipe container and working fluid and neglected the axial heat conduction. The numerical model uses an iterative mathematical process involving the following problems:

1. Heat transfer in the evaporating film on a rough surface
2. Heat transfer in the condensate film on the fin top surface
3. Heat conduction in a metallic fin and liquid meniscus
4. Fluid circulation in the AGHPs

The first three problems were detailed in Khrustalev and Faghri (1995b), and their interactions with one another, as well as the fluid circulation analyses, are described by Khrustalev and Faghri (1995a).

A comprehensive review of advanced modeling efforts of flat miniature heat pipes with axial capillary grooves was presented by Faghri and Khrustalev (1997). Lefèvre et al. (2010) extended the model by Khrustaler and Faghri (1995a) by approximating the axial temperature distribution of a heat pipe using variable thermal resistances as a nodal model. It was concluded that the thermal resistances of a heat pipe may be significant when the heat pipe wall is thick and/or the operating temperature is in the intermediate to high range. Do et al. (2008) used the models formulated by Khrustalev and Faghri (1995a, 1995b) with inclusion of the axial heat conduction in the heat pipe wall. Figure 5.11a shows the maximum heat transport rate as predicted by Do et al. (2008) compared to the experimental data and predictions of the theoretical model of Hopkins et al. (1999). The wall temperature versus the heat input at different locations is presented in Figure 5.11b.

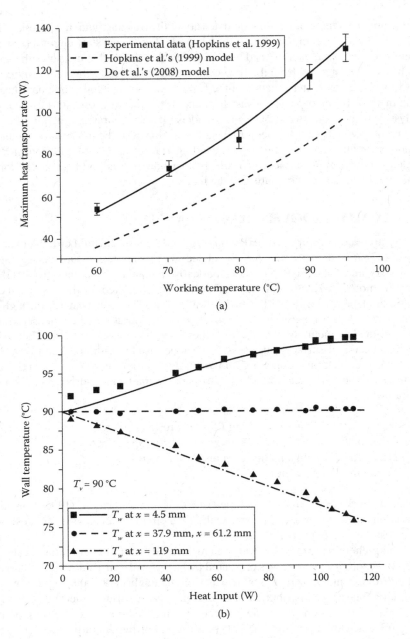

FIGURE 5.11 Comparison of the model predictions with (symbols) experimental data from Hopkins et al. (1999) and (lines) numerical simulation results from the Do et al. (2008) model: (a) maximum heat transport rate and (b) wall temperature. (Reprinted from *International Journal of Heat and Mass Transfer*, 51, Do, K. H., Kim, S. J. and Garimella, S. V., A Mathematical Model for Analyzing the Thermal Characteristics of a Flat Micro Heat Pipe with a Grooved Wick, 4637–4650, Copyright (2008), with permission from Elsevier.)

5.8 THERMOSYPHON ANALYSIS

Early analyses of thermosyphons were mainly focused on the condenser section. However, two main approaches have evolved for the full numerical simulation of entire two-phase closed wickless thermosyphons including both condenser and evaporator sections. The different approaches are defined by their definitions of the liquid and vapor phases. The separated flow model approach discusses the liquid and vapor phases and the corresponding mass transfer separately. The multi-fluid Eulerian approach allows for liquid and vapor penetration. Alizadehdakhel et al. (2010) and Fadhl et al. (2013) outlined the multi-fluid Eulerian approach using volume of fluid techniques presented by Faghri and Zhang (2006). Despite showing promise, the multi-fluid Eulerian approach requires further effort to draw more realistic results with independence of any empirical constant.

Spendel (1984) modeled the condenser section of a conventional thermo-syphon with an incompressible, two-dimensional vapor domain. The study included a parametric effort in which the effect of the interfacial shear stress and vapor pressure drop on the falling film thickness was described. Spendel (1984) determined that the localized Nusselt number may vary up to 60% from the approximation originally described by Nusselt for the condenser section of a thermosyphon.

Faghri et al. (1989) examined the heat transfer of conventional and concentric annular thermosyphons and developed an improved prediction for the flooding limit. An analysis of the empirically derived interfacial shear stress of the counter-flowing vapor on the falling liquid film was included for the condenser section of thermosyphons.

A transient, two-dimensional model with variable properties was presented by Harley and Faghri (1994b). Their study used a generalized quasi-steady Nusselt-type solution for the falling film coupled with the complete two-dimensional solution of the vapor flow. This model was particularly significant in the description of thermosyphon behavior and performance because it included the entire thermosyphon, instead of only focusing on the condenser section, as was the case in previous work. The vapor flow was simulated using a two-dimensional transient formulation, which was then coupled with the unsteady heat conduction in the pipe wall. Furthermore, the quasi-steady falling con-densate film in each of the condenser, adiabatic, and evaporator sections was modeled using the variable vapor condensation rate, interfacial shear stress, and vapor pressure drop.

Harley and Faghri (1994a) used an extension of their previous model (Harley and Faghri 1994b) for the analysis of gas-loaded thermosyphons. As before, the model used a generalized quasi-steady Nusselt-type solution for the falling film coupled with the complete two-dimensional solution of the vapor flow. The extension of the model is particularly important because the entire thermosyphon was in the domain of the model, unlike previous works that focused solely on the condenser section.

A numerical model by Zuo and Gunnerson (1995) presented the effects of inclination angle on thermosyphon performance. The liquid–vapor interfacial shear stress and the effect of working fluid charge were examined for various inclination angles. Zuo and Gunnerson (1995) also indicated the mechanisms that ultimately cause the dry out and flooding limitations of an inclined thermosyphon. Lin and Faghri (1997b) formulated a one-dimensional, steady-state mathematical model for the natural circulation of two-phase flow in a thermosyphon with a tube separator. Liquid fill ratios were considered for steady-state operation, and the distribution of void fractions along the thermosyphon was obtained for various operating conditions. Lin and Faghri (1998b) later studied the hydrodynamic stability of the natural circulation of two-phase flow in a high-performance thermosyphon with a tube separator using numerical analysis. Results indicated that flow instability may be significantly affected by operational parameters, including the operating temperature, heat rate, and thermosyphon inclination angle.

El-Genk and Saber (1999) developed a one-dimensional numerical model for the description of a functional range (or operation-envelope) for closed, two-phase thermosyphons at steady state. They concluded that increased thermosyphon diameter, evaporator length, or vapor temperature expanded the functional range of the thermosyphon, though an increase in the length of the condenser or adiabatic section resulted in only minor increases in the envelope's upper and lower boundaries. Pan (2001) accounted for the interfacial shear stress due to mass transfer and interfacial velocity for condensation in a two-phase closed thermosyphon. Both the relative velocity ratio and the momentum transfer were identified as significant factors for the condensation heat transfer, and a sub-flooding limit was proposed for predicting the interaction between condensation and evaporation in a thermosyphon.

Jiao et al. (2012) developed a predictive model for the effect of the filling ratio on the dryout, flooding, and boiling limits. Experimental tests were conducted using nitrogen as the working fluid, and a comparison between the experimental and theoretical results was presented. The maximum filling ratio was introduced, which determines an upper limit for the heat transfer performance in a thermosyphon.

Shabgard et al. (2014) investigated the performance of a thermosyphon under various fill conditions, including overfilled, optimum-fill, and underfilled. Simulation indicated that the evaporator temperature of an underfilled thermosyphon rapidly increases due to the onset of dryout conditions. The optimally filled thermosyphon had the shortest response time and lowest thermal resistance (Figure 5.12). However, a slight increase in the power input caused a breakdown of the condensate film. The overfilled thermosyphon yielded a slightly slower thermal response and had a larger thermal response than for the optimal fill condition, but it provided a more stable condensate film. Therefore, Shabgard et al. recommended that an optimally filled thermosyphon should be filled with a small amount of additional working fluid to prevent breakdown of the liquid film and to ensure both optimal and stable steady-state operations.

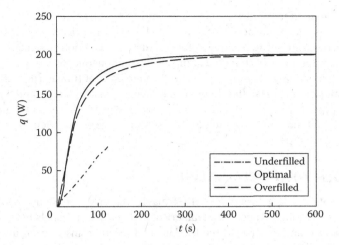

FIGURE 5.12 Temporal variations of the output heat from the condenser section for the thermosyphon with various filling ratios. (Reprinted from *International Journal of Heat and Mass Transfer*, Vol. 70, Shabgard, H., Xiao, B., Faghri, A., Gupta, R. and Weissman, W., Thermal Characteristics of a Closed Thermosyphon Under Various Filling Conditions, 91–102, Copyright (2014), with permission from Elsevier.)

5.9 ROTATING HEAT PIPE ANALYSIS

Gray (1969) first proposed the concept of a rotating heat pipe and showed that the design is capable of transporting significantly more heat than a conventional stationary heat pipe. Faghri et al. (1993) analyzed the vapor flow of a rotating heat pipe using a two-dimensional axisymmetric model. Their study examined the influence of rotation rate, vapor pressure drop, and interfacial shear stress. A detailed transient numerical simulation was developed by Harley and Faghri (1995) for rotating heat pipes that included the thin liquid condensate film as well as the vapor flow and heat conduction in the pipe wall. Lin and Faghri (1997a) examined the mechanical and thermal behavior of axially rotating heat pipes in the stratified flow regime with both cylindrical and stepped-wall configurations that agreed well with existing experimental data. Lin and Faghri (1997c) also presented a related model of a miniature heat pipe with a grooved inner wall surface. Influence on the hydrodynamic performance was analyzed based on changes to operating temperature, rotational speed, and liquid–vapor interfacial shear stress. The maximum performance and optimal liquid fill were discussed, and the pressure drops of the axial liquid and vapor flows were presented.

The hysteresis behavior of a rotating heat pipe with a stepped-wall geometry in the annular flow regime was discussed by Lin and Faghri (1998a). A theoretical model was formulated by Lin and Faghri (1999) for a rotating miniature heat pipe with an axial triangular-grooved internal surface. The formulation used thin liquid film vaporization heat transfer theory for prediction of the evaporative heat transfer in the micro region. Furthermore, the effects of disjoining pressure, surface tension, and centrifugal force on the flow were described. The influence of rotational speed was deemed negligible on the evaporation heat transfer in the micro region and could be neglected. Cao (2010) numerically and analytically studied miniature high-temperature rotating heat pipes for gas turbine and cooling vane applications.

Harley and Faghri (2000) included the effects of noncondensable gases in a rotating heat pipe core, and they used a conjugate coupled heat transfer model to account for the vapor pressure drop, thin liquid film, interfacial shear stress, and heat conduction in the pipe. A parametric study was reported with emphasis on the influence of rotational speed, heat input/output, and the mass of the noncondensable gas within the rotating heat pipe.

5.10 LOOP HEAT PIPE ANALYSIS

A full description of loop heat pipe (LHP) behavior requires thermal and hydrodynamic couplings among their components. Numerical and analytical models have been developed to describe and understand the fundamental mechanisms of each interaction, as well as to quantitatively predict LHP operational characteristics. In most investigations, energy and momentum balances are used to model the thermal and hydrodynamic phenomena within LHPs. Some notable steady-state modeling efforts include the works by Maydanik et al. (1994), Kaya and Hoang (1999), Hoang and Kaya (1999), Kaya and Ku (1999), Muraoka et al. (2001), Chuang (2003), Hamdan (2003), Kaya and Ku (2003), Furukawa (2006), Kaya and Goldak (2006), and Launay et al. (2007c, 2008).

Launay et al. (2008) presented separate closed-form solutions for LHPs operating in the variable conductance mode (VCM) and the fixed conductance mode (FCM). The two distinct modes of LHP operation may be indicated by the operational curve obtained by comparing the heat input with the operating temperature of the LHP (typically the compensation chamber temperature or evaporator temperature). The models also described the distribution and function of the condenser heat transfer area: one part is used for condensation, while the other accounts for liquid subcooling.

The LHP operating temperature was predicted by Launay et al. (2008) as a function of the heat input, as shown in Figures 5.13 and 5.14, for both acetone- and ammonia-filled loop heat pipes, respectively. As shown in Figures 5.13 and 5.14, the full numerical simulation and simplified closed-form solutions are in agreement with the experimental measurements for two LHP designs (Chuang 2003; Boo and Chung 2004). The relative difference between the full numerical simulation and closed-form solution is at a maximum within the region of transition between VCM and FCM operation, though the maximum difference is less than 15%. The general agreement between the numerical, closed-form analytical and experimental results validates the use of both the model and closed-form methods as tools for future LHP design.

Transient LHP publications include Cullimore and Bauman (2000), Hoang and Ku (2003), Launay et al. (2007a, 2007b), Kaya et al. (2008), Chernysheva and Maydanik (2008), and Khrustalev (2010). An innovative model for the transient analysis of two-phase LHP systems and capillary evaporators was conducted by Cullimore and Bauman (2000). The analysis was later extended by Khrustalev (2010) to intricate loop heat pipe systems involving complex radiators, multiple components, and various environmental conditions. Ambirajan et al. (2012) comprehensively reviewed the fundamentals, operation, and design of loop heat pipes.

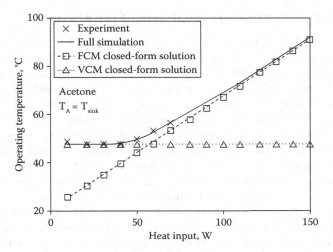

FIGURE 5.13 Comparison of LHP modeling predictions (From Launay, S., et al., *J. Thermophys. Heat Transfer*, 22(4), 623–631, 2008. With permission.) with experimental results (From Boo, J.H. & Chung, W.B., *Proceedings of the 13th International Heat Pipe Conference*, Shanghai, China, pp. 259–264, 2004. With permission.) using acetone as the working fluid. (From Analytical Model for Characterization of Loop Heat Pipes, Launay, S., Sartre, V. & Bonjour, J., 2008, reprinted by permission of the American Institute of Aeronautics and Astronautics.)

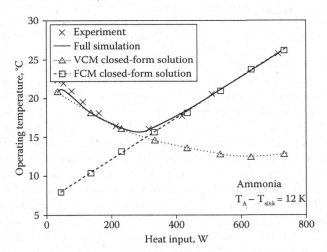

FIGURE 5.14 Comparison of LHP modeling predictions (From Analytical Model for Characterization of Loop Heat Pipes, Launay, S., Sartre, V. and Bonjour, J., 2008, reprinted by permission of the American Institute of Aeronautics and Astronautics.) with experimental results (From Chuang, P.A., An Improved Steady-State Model of Loop Heat Pipe Based on Experimental and Theoretical Analyses. Ph.D. Thesis, Pennsylvania State University, State College, PA, 2003) with ammonia as the working fluid. (From Analytical Model for Characterization of Loop Heat Pipes, Launay, S., Sartre, V. & Bonjour, J., 2008, reprinted by permission of the American Institute of Aeronautics and Astronautics.)

5.11   CAPILLARY PUMPED LOOP HEAT PIPE ANALYSIS

In general, capillary pumped loop heat pipes (CPLs) may be analyzed using three different approaches, with varying degrees of complexity and accuracy. The first approach was used by Kiper et al. (1990) and calculated the heat transfer of the CPL without inclusion of fluid flow in the wick or vapor regions of the evaporator section. Furthermore, the solution method used a lumped analysis of the evaporator and subcooler with an assumed exponential temperature profile. This analytical approach utilized a number of approximations to solve the energy equations and, without the inclusion of a coupled fluid flow and heat transfer analysis in the evaporator, no meaningful conclusions could be made. The second type of approach includes a detailed and accurate analysis of the evaporator component for both steady and transient operating conditions and is made possible by the solution of the complete differential forms of the momentum and energy equations (Cao and Faghri 1994b, 1994c). This methodology may be considered as the most suitable for determination of CPL system behavior. A third approach has been used in which semi-empirical correlations are used for the calculation of pressure drop and heat transfer coefficients in different parts of the CPL under 1 g conditions (Kroliczek et al. 1984; Ku et al. 1986a, 1986b, 1987a, 1987b, 1988, 1993; Chalmers et al. 1988; Benner et al. 1989; Schweickart and Buchko 1991). The complex transient start-up phenomena in a CPL have been considered by Cullimore (1991).

5.12   MICRO HEAT PIPE ANALYSIS

A detailed description of the heat and mass transfer processes in a micro heat pipe (MHP) was developed by Khrustalev and Faghri (1994). Both the liquid distribution and the corresponding thermal characteristics were given as functions of the liquid charge and the applied heat load. In addition, the liquid flow in an MHP with a polygonal cross section was considered. The model accounted for the liquid flow in the triangular-shaped corners using the variation of the curvature of the free liquid surface and the interfacial shear stresses due to liquid–vapor interaction. The maximum heat transfer capacity and thermal resistance of the MHP were predicted and compared to experimental data, and the importance of the liquid fill, minimum wetting contact angle, and shear stresses at the liquid–vapor interface was demonstrated.

Several extended models have been used to improve the original model of Khrustalev and Faghri (1994). Sartre et al. (2000) and Suman and Kumar (2005) included heat conduction in the wall for a similar analysis of MHPs with polygonal cross-sectional areas. Wang and Peterson (2002) and Launay et al. (2004) simulated arrays of MHPs made from several aluminum wires bound between two aluminum sheets.

5.13   PULSATING HEAT PIPE ANALYSIS

There are two main types of pulsating heat pipes (PHPs), looped (Figure 5.15a) and unlooped (Figure 5.15b), which are classified specifically by whether or not the two ends of the PHP connect. PHP modeling differs from that of conventional heat pipes primarily due to the characteristic slug flow patterns during operation. Miyazaki and

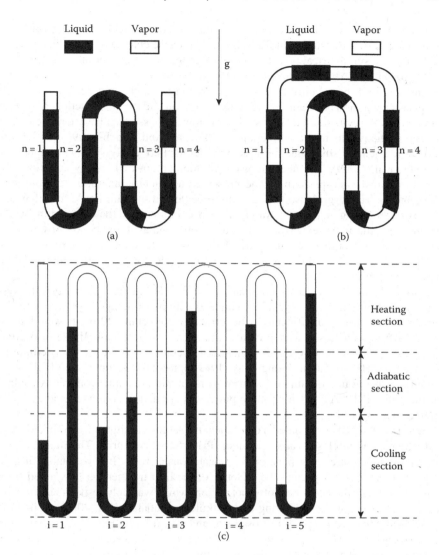

FIGURE 5.15   Pulsating heat pipes: (a) open-loop, (b) closed-loop (Reprinted from *Journal of Heat Transfer*, Shafii, M.B., Faghri, A. and Zhang, Y., Thermal Modeling of Unlooped and Looped Pulsating Heat Pipes, 2001, with permission from ASME.), and (c) open-loop PHP with arbitrary turns ($N = 5$) (Analytical Model for Characterization of Loop Heat Pipes, Launay, S., Sartre, V. and Bonjour, J., 2008, reprinted by permission of the American Institute of Aeronautics and Astronautics.)

Akachi (1996) gave a simple analytical model for the self-exciting oscillation that was observed during experimentation. It was noted that the oscillation of the void fraction caused a resulting change in the heat transfer rate, and the effect on the reciprocal excitation of pressure oscillations was studied. An optimal charge ratio for a particular PHP was presented. At charge ratios that are too high, the PHP experiences a gradual increase in pressure and a subsequent sudden drop in pressure. In contrast, an insufficient fluid charge will cause chaotic pressure fluctuations instead of the symmetrical pressure wave generated at an appropriate charge ratio.

Miyazaki and Akachi (1998) later derived the wave equation of pressure oscillations in a PHP based on the self-excited oscillation and again accounted for the reciprocal excitation between the pressure oscillation and the void fraction. Miyazaki and Arikawa (1999) studied the oscillatory flow and associated wave velocity and compared the results to the predictions of Miyazaki and Akachi (1996), with fairly good agreement. A simplified numerical model for PHP analysis was developed by Hosoda et al. (1999). The analysis used the momentum and energy equations for two-dimensional two-phase flow and presented transient temperature and pressure distributions. However, Hosoda et al. (1999) neglected the thin liquid film that surrounds a vapor plug on the tube wall and the friction between the tube and the working fluid. Furthermore, experimental results were used as initial conditions, and the numerical results of the pressure in the PHP were higher than the experimental values. Despite this, the model showed that the propagation of vapor plugs induced fluid flow in the capillary tubes.

Zuo et al. (1999, 2001) adopted an approach for PHP modeling in which the system was simplified to an equivalent single spring-mass-damper system, with the parameters of the system affected by heat transfer. Zuo et al. (2001) showed that the spring stiffness increased with time and that, as a result, the amplitude of oscillation must decrease. However, this conclusion is contrary to the steady oscillations observed in PHP operation. Wong et al. (1999) considered an open-loop PHP using a multiple spring-mass-damper system with adiabatic boundary conditions applied for the entire PHP. A sudden pressure pulse was applied to simulate localized heat input into a vapor plug.

Shafii et al. (2001) formulated a theoretical model for the simulation of liquid slugs and vapor plugs in closed- and open-loop PHPs with two turns. The model calculated the temperature, plug position, and heat transfer rates. It was notable that, in contrast to what was expected, the majority of the heat transfer (95%) resulted from sensible heat, not the latent heat of vaporization, which was only responsible for driving and maintaining the oscillating flow. Sakulchangsatjatai et al. (2004) applied the model from Shafii et al. (2001) to closed-end and closed-loop PHPs. The investigation approximated the PHP behavior using oscillating two-phase heat and mass transfer in a straight pipe and neglected the thin liquid film between the vapor plug and pipe wall.

Zhang et al. (2002) developed an analytical model for the investigation of oscillatory flow in a U-shaped miniature channel (a building block of PHPs). Most significantly, and in contrast to other mathematical models, the governing equations were nondimensionalized. Flow in the tube was reduced to a description using two dimensionless parameters, a nondimensional temperature difference, and the evaporation and condensation heat transfer coefficients. The initial displacement of the liquid slug and gravity was found to have no effect on the amplitude and angular frequency of the oscillation. Instead, the amplitude and angular frequency of the oscillation were increased by increasing the dimensionless temperature difference. A correlation was presented for the dependence of the amplitude and angular frequency of the oscillations of the heat transfer coefficients to the temperature difference.

Zhang and Faghri (2003) investigated the oscillatory flow of a closed-end, pulsating heat pipe with an arbitrary number of turns (Figure 5.15c). It was found that a PHP with only a few turns (less than six) had an amplitude and oscillation frequency that were independent of the number of turns. For the PHP with only a few turns, the motion of the vapor plugs was identical for odd-numbered plugs at steady state. Even-numbered plugs also demonstrated identical motion. Both odd- and even-numbered plugs also had identical amplitude; however, they were out of phase by a factor of π. For a PHP of greater than six turns, the odd- and even-numbered plugs did not demonstrate identical oscillation, and each lagged slightly behind the next. Each plug remained out of phase by π.

Dobson and Harms (1999) studied a PHP with two parallel open ends that were aligned in the same direction and an applied heat flux at the evaporator section, which caused oscillatory fluid motion and a net thrust. A corresponding numerical solution of the energy equation and the equation of motion for a vapor plug was presented to predict the vapor plug's temperature, position, and velocity. Dobson (2004, 2005) discussed the use of an open-ended PHP with two check valves as a water pump. However, the proposed mass flow rates were not sufficient for most practical applications. An improved analysis for liquid slug oscillation was presented that described the pressure difference, friction, gravity, and surface tension.

Zhang and Faghri (2002) proposed a method for the modeling of a PHP with one open end by analyzing thin film evaporation and condensation. The heat transfer in the evaporator and condenser sections was calculated as the sum of the evaporative (or condensation) heat transfer in the thin liquid film and at the meniscus, and the sensible heat transfer to the liquid slug was included. Results indicated that the overall heat transfer was dominated by the transfer of sensible heat—not latent heat. Shafii et al. (2002) extended the previous numerical model from Shafii et al. (2001) to include an analysis of evaporation and condensation heat transfer in the thin liquid film between the liquid and vapor plugs. Figure 5.16 shows that the dominant mechanism of heat transfer was due to sensible heat transfer (~95%) for both open- and closed-PHPs. It was also noted that increased heat transfer was obtained for a corresponding increase in surface tension of the working fluid and an increased diameter of the tube, while lower heat transfer correlated to a lower wall temperature of the heated section.

A mathematical model was presented by Liang and Ma (2004) that described the oscillatory characteristics of slug flow in a capillary tube. Numerical results were also presented that indicated that the isentropic bulk modulus generated stronger oscillations than the isothermal bulk modulus. Among the parameters tested, it was found that the capillary force, gravitational force, and initial pressure distribution of the working fluid most strongly affected the frequency and amplitude of the oscillatory motion. Ma et al. (2002, 2006) developed an analytical model that described the oscillations of a liquid slug using force balances, including the thermally driven, capillary, frictional, and elastic restoring forces. Ma et al. (2006) concluded that the oscillating motion of the slug was dependent on the charge ratio, total characteristic length, diameter, temperature difference between the evaporator and condenser, and the working fluid and the operating temperatures. However, the mathematical model was found to under-predict the

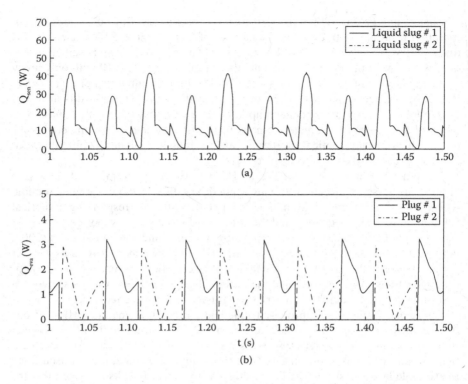

FIGURE 5.16 Heat transfer rate: (a) sensible heat and (b) evaporative heat. (From Shafii, M.B., et al., *Int. J. Numer. Methods Heat Fluid Flow*, 12(5), 585–609, 2002. With permission.)

driving temperature difference between the evaporator and condenser compared to the existing experimental results (Ma et al., 2002).

Holley and Faghri (2005) formulated a numerical model for PHP with a sintered copper capillary wick and with flow channels of various diameters. It was noted that, when the channel was of smaller diameter, it more readily induced circulation of the working fluid and thus increased the heat load capability of the PHP. The model predicted better performance of the PHP in the bottom heating mode (a more uniform temperature) than for the top heating mode, and it was observed that a greater number of parallel channels increased the heat capacity and decreased the sensitivity of the PHP to gravity.

An artificial neural network (ANN) was used by Khandekar et al. (2002) to predict PHP behavior and performance. The ANN had a fully connected feed-forward configuration and was trained using 52 sets of experimental data from a closed-loop PHP. The heat input and fill ratio of each experimental data set were system inputs, and the model predicted the effective thermal resistance of the PHP. However, many operational parameters were neglected, including tube diameter, number of parallel channels, length of the PHP, inclination angle, and the working fluid properties. The considerable number of inputs required for effective thermal modeling using the ANN approach makes it highly dependent on experimental data.

Khandekar and Gupta (2007) used a commercial code (FLUENT) to model the heat transfer of an embedded PHP in a radiator plate. The oscillatory flow and heat transfer of the PHP was not modeled in this investigation, and the contribution of the PHP was only considered using an effective thermal conductivity that was obtained experimentally. Zhang and Faghri (2008) comprehensively reviewed pulsating heat pipe analysis, and Khandekar et al. (2010) presented experimental and theoretical methods for prediction of the hydrodynamics of unidirectional two-phase Taylor bubble flows, and they recommended areas of further research for the modeling of pulsating heat pipes.

5.14 INTEGRATED HEAT PIPES AND PHASE CHANGE MATERIALS

The highly effective thermal conductivities of heat pipes and thermosyphons have promoted their use in many energy-related applications, particularly in latent heat thermal energy storage (LHTES) systems. Utilization of heat pipes is an attractive solution for reduction of the relatively high thermal resistances of phase change materials (PCMs). Sharifi et al. (2012) used numerical simulation to model the melting of a PCM that was contained in a vertical cylindrical enclosure and subjected to a radial heat flux by an embedded heat pipe. The melting rate of the PCM-heat pipe system was compared to that of a system heated by an isothermal cylindrical surface, a solid rod, and a hollow tube (all with the same exterior dimensions). Melting effectiveness was defined as the ratio of the PCM liquid fraction for each system compared to the liquid fraction of the solid rod-enhanced system. The time-dependent melting effectiveness for the isothermal condition, heat pipe, and tube case study is illustrated in Figure 5.17. It should be noted that the effectiveness for the heat pipe-enhanced PCM is above unity, indicating improved heat transfer performance relative to the rod-enhanced system. The ideal case of heat transfer performance is that attained for

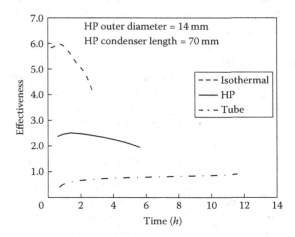

FIGURE 5.17 Effectiveness histories for isothermal, heat pipe, and tube cases. (Reprinted from *International Journal of Heat and Mass Transfer*, 55, Sharifi, N., Wang, S., Bergman, T., and Faghri, A., Heat Pipe-assisted Melting of a Phase Change Material, 3458–3469, Copyright (2012), with permission from Elsevier.)

an isothermal surface with a temperature identical to that of the heat pipe evaporator, and the lower limit of performance is that obtained for simple conduction through a hollow tube.

A one-dimensional thermal network approach for modeling and prediction of heat transfer in a latent heat pipe thermal energy storage system for solar thermal electricity generation was studied by Shabgard et al. (2010). The thermal network was developed for a system using the multiple heat pipes between a heat transfer fluid and a PCM. Analysis included an LHTES system with the heat transfer fluid flowing through a hollow pipe with a surrounding PCM. A separate LHTES configuration involved an internally loaded PCM within the tube in cross flow with the heat transfer fluid. Shabgard et al. (2010) included the effects of natural convection in the thermal network, using potassium nitrate (KNO_3) as the PCM and stainless-steel heat pipes with mercury as the internal working fluid. It was noted that the heat pipe-enhanced LHTES systems improved thermal performance for both melting and solidification processes (the heat transfer fluid acting as a heat source and sink, respectively). The charging (melting) effectiveness was increased by approximately 60%, and the discharging (solidification) effectiveness improved by 40%.

In a numerical investigation by Nithyanandam and Pitchumani (2013), an LHTES system was considered for concentrating solar power generation using integrated thermosyphons to enhance the thermal conductivity of the PCMs. A transient, computational method was used to better understand the effect of design configuration and thermosyphon arrangement for both charging and discharging of the system. Configurations of improved effectiveness and energy storage/retrieval rate were presented and considered both in terms of the per unit cost of the thermosyphons as well as in comparison to the performance of equivalent systems using embedded heat pipes. The PCM under consideration was KNO_3, and the stainless-steel thermosyphons used Therminol VP-1 as the internal working fluid. Similar to the study by Shabgard et al., (2010), two configurations were included in the analysis, one with the PCM concentrically loaded around the exterior of a tube containing an internal flow of heat transfer fluid and another with the PCM housed within the tube with an external cross-flow of heat transfer fluid. It was determined that faster solidification and melting rates were obtained using embedded thermosyphons. Furthermore, it was noted that a higher power density was possible in systems utilizing embedded heat pipes, but the highest power density per unit cost for both charging and discharging was obtained using the proposed embedded thermosyphon configuration.

5.15   HEAT PIPES WITH NANOFLUIDS AND NANOPARTICLES

Integration of heat pipe working fluids with nanoparticles and nanofluids provides the potential for improved thermal performance over conventional heat pipes. Shafahi et al. (2010) used a two-dimensional, steady-state analysis with an incompressible vapor flow to obtain the velocity, pressure, temperature, and maximum heat transfer limit for a conventional, cylindrical heat pipe and porous wick charged with aqueous aluminum oxide (Al_2O_3), copper oxide (CuO), and titanium dioxide (TiO_2) nanoparticles. The wall temperature was assumed uniform in the condenser and evaporator, and the temperature distribution for the evaporator, adiabatic, and condenser sections was modeled using previously developed analytical methods. It was observed that the liquid velocity decreased with increased nanoparticle concentration. In addition, the heat transfer of the heat pipe increased, along with the concentration of nanoparticles, and the optimum concentration of each nanofluid was found. Do and Jang (2010) studied nanofluid enhancement in a flat MHP and formulated two different mathematical models under the assumptions of one-dimensional steady incompressible axial flow, one-dimensional axial temperature difference in the heat pipe wall, and negligible convection in the liquid and vapor phases. The heat transfer and thermal resistance of the heat pipe with respect to the enhanced effective thermal conductivity of the working fluid and the increased heat transfer and surface area of the conductive deposition of nanoparticles surrounding the grooved wick structure were solved for in each model. It was concluded that the thermal resistance of the nanofluid-charged heat pipe is dependent on two opposing phenomena. The thin deposition layer of nanoparticles in the evaporator caused an increased liquid pressure drop within the nanoparticle layer, which limited evaporative heat transfer. In contrast, the deposition of nanoparticles results in an additional highly conductive surface area of the evaporator.

Ghanbarpour and Khodabandeh (2015) developed a steady-state analytical model alongside their experimental study to predict the thermal conductivity, viscosity, entropy generation, and thermal resistance of a nanofluid- and basefluid-charged heat pipe using a thermal network approach. The model assumed negligible axial thermal resistances of the heat pipe wall, wick, and vapor flow. The experiment indicated a reduction in entropy generation between 3 and 13.5% when using nanofluids at concentrations of 1 to 5 vol.%, and the analytical model predicted a reduction of entropy generation in reasonable agreement with the experimental data. Mashaei and Shahryari (2015) modified the analytical solution of the energy equation from Shabgard and Faghri (2011) using separation of variables to predict the wall temperature of a conventional cylindrical heat pipe charged with water-based Al_2O_3 and TiO_2 nanofluids. The analysis assumed steady-state conditions; constant thermal conductivities of the nanoparticles, basefluid, wall, and solid phase of the wick; and a fully saturated wick. It was also assumed that the heat transfer in the porous wick mainly took place via conduction. The analytical model predicted that higher concentrations of nanoparticles or nanoparticles of smaller diameter increased the heat transfer coefficient between the pipe wall and saturated vapor flow and decreased the temperature difference between the evaporator and condenser sections of the heat pipe.

5.16 CONCLUSION

An overview of analysis and numerical simulations of different types of heat pipes under various operating conditions is presented in this review. The significant and rapid progression of heat pipe technology is examined from a perspective of the state-of-the-art modeling and the full simulation that has been made possible in the last few decades. Simulations are able to accurately predict the thermal performance of heat pipes under various operating conditions, including steady-state, continuum transient, and frozen start-up solutions, despite the associated complex multiphase and multidomain transport phenomena. Steady-state and transient heat pipe simulations involve conjugate heat transfer with the wall, wick, and vapor, though pulsating and loop heat pipes require more fundamental research efforts to explain the physical phenomena of these devices. Furthermore, simulation of the liquid–vapor interface requires consideration of the multiphase phenomena within various wick structures and will provide future understanding and prediction of the heat transport limitation in heat pipes.

REFERENCES

Aghvami, M. & Faghri, A., 2011. Analysis of Flat Heat Pipes with Various Heating and Cooling Configurations. *Applied Thermal Engineering*, 31(14–15), 2645–2655. DOI: 10.1016/j.applthermaleng.2011.04.034.

Alizadehdakhel, A., Rahimi, M. & Alsaira, A., 2010. CFD Modeling of Flow and Heat Transfer in a Thermosyphon. *INTL Communications of Heat and Mass Transfer*, 37, 312–318.

Ambirajan, A., Adoni, A. A., Vaidya, J. S., Rajendran, A. A., Kumar, D. & Dutta, P., 2012. Loop Heat Pipes: A Review of Fundamentals, Operation, and Design. *Heat Transfer Engineering*, 33(4–5), 387–405. DOI: 10.1080/01457632.2012.614148.

Benner, S., Costello, F. & Ku, J., 1989. SINFAC Simulation of a High-Power Hybrid CPL. *AIAA-89-0316, Proceedings of the 27th Aerospace Sciences Meeting*, Reno, NV.

Boo, J. H. & Chung, W. B., 2004. Thermal Performance of a Small-Scale Loop Heat Pipe with PP Wick. *Proceedings of the 13th International Heat Pipe Conference*, Shanghai, China, pp. 259–264.

Cao, Y., 2010. Miniature High-Temperature Rotating Heat Pipes and their Applications in Gas Turbine Cooling. *Frontiers in Heat Pipes (FHP)*, DOI: 10.5098/fhp.v1.2.3002.

Cao, Y. & Faghri, A., 1990. Transient Two-Dimensional Compressible Analysis for High-Temperature Heat Pipes with Pulsed Heat Input. *Numerical Heat Transfer; Part A: Applications*, 18(4), 483–502. DOI: 10.1080/10407789008944804.

Cao, Y. & Faghri, A., 1991. Transient Multidimensional Analysis of Nonconventional Heat Pipes with Uniform and Nonuniform Heat Distributions. *Journal of Heat Transfer*, 113(4), 995–1002. DOI: 10.1115/1.2911233.

Cao, Y. & Faghri, A., 1992. Closed-Form Analytical Solutions of High-Temperature Heat Pipe Startup and Frozen Startup Limitation. *Journal of Heat Transfer*, 114(4), 1028–1035. DOI: 10.1115/1.2911873.

Cao, Y. & Faghri, A., 1993a. Conjugate Modeling of High-Temperature Nosecap and Wing Leading-Edge Heat Pipes. *Journal of Heat Transfer*, 115(3), 819–822. DOI: 10.1115/1.2910765.

Cao, Y. & Faghri, A., 1993b. Simulation of the Early Startup Period of High-Temperature Heat Pipes from the Frozen State by a Rarefied Vapor Self-Diffusion Model. *Journal of Heat Transfer*, 115(1), 239–246. DOI: 10.1115/1.2910655.

Cao, Y. & Faghri, A., 1993c. A Numerical Analysis of High-Temperature Heat Pipe Startup from the Frozen State. *Journal of Heat Transfer*, 115(1), 247–254. DOI: 10.1115/1.2910657.

Cao, Y. & Faghri, A., 1994a. Analytical Solutions of Flow and Heat Transfer in a Porous Structure with Partial Heating and Evaporation on the Upper Surface. *International Journal of Heat and Mass Transfer*, 37(10), 1525–1533. DOI: 10.1016/0017-9310(94)90154-6.

Cao, Y. & Faghri, A., 1994b. Conjugate Analysis of a Flat-Plate Type Evaporator for Capillary Pumped Loops with Three-Dimensional Vapor Flow in the Groove. *International Journal of Heat and Mass Transfer*, 37(3), 401–409. DOI: 10.1016/0017-9310(94)90074-4.

Cao, Y., Faghri, A. & Mahefkey, E. T., 1989. The Thermal Performance of Heat Pipes with Localized Heat Input. *International Journal of Heat and Mass Transfer,* 32(7), 1279–1287. DOI: 10.1016/0017-9310(89)90028-8.

Chalmers, D. R., Fredley, J., Ku, J. & Kroliczek, E. J., 1988. Design of a Two-Phase Capillary Pumped Flight Experiment, SAE 88-1086. *Proceedings of 18th Intersociety Conference on Environmental Systems,* San Francisco, CA.

Chen, M. M. & Faghri, A., 1990. An Analysis of the Vapor Flow and the Heat Conduction through the Liquid-Wick and Pipe Wall in a Heat Pipe with Single Or Multiple Heat Sources. *International Journal of Heat and Mass Transfer,* 33(9), 1945–1955. DOI: 10.1016/0017-9310(90)90226-K.

Chernysheva, M. A. & Maydanik, Y. F., 2008. Numerical Simulation of Transient Heat and Mass Transfer in a Cylindrical Evaporator of a Loop Heat Pipe. *International Journal of Heat and Mass Transfer,* 51(17–18), 4204–4215. DOI: 10.1016/j.ijheatmasstransfer.2007.12.021.

Chuang, P. A., 2003. An Improved Steady-State Model of Loop Heat Pipe Based on Experimental and Theoretical Analyses. Ph.D. Thesis, Pennsylvania State University, State College, PA.

Cullimore, B., 1991. *Startup Transient in Capillary Pumped Loops.* AIAA-91-1374, Honolulu, HI.

Cullimore, B. & Bauman, J., 2000. Steady State and Transient Loop Heat Pipe Modeling. SAE 2000-01-2316, *Proceedings of the 34th International Conference on Environmental Systems (ICES),* Toulouse, France.

Do, K. & Jang, S., 2010. Effect of Nanofluids on the Thermal Performance of a Flat Micro Heat Pipe with a Rectangular Grooved Wick. *International Journal of Heat and Mass Transfer,* 53, 2183–2192. DOI: 10.1016/j.ijheatmasstransfer.2009.12.020.

Do, K. H., Kim, S. J. & Garimella, S. V., 2008. A Mathematical Model for Analyzing the Thermal Characteristics of a Flat Micro Heat Pipe with a Grooved Wick. *International Journal of Heat and Mass Transfer,* 51(19–20), 4637–4650. DOI: 10.1016/j.ijheatmasstransfer.2008.02.039.

Dobson, R. T., 2004. Theoretical and Experimental Modelling of an Open Oscillatory Heat Pipe Including Gravity. *International Journal of Thermal Sciences,* 43(2), 113–119. DOI: 10.1016/j.ijthermalsci.2003.05.003.

Dobson, R. T., 2005. An Open Oscillatory Heat Pipe Water Pump. *Applied Thermal Engineering,* 25(4), 603–621. DOI: 10.1016/j.applthermaleng.2004.07.005.

Dobson, R. T. & Harms, T. M., 1999. Lumped Parameter Analysis of Closed and Open Oscillatory Heat Pipes. *Proceedings of the 11th International Heat Pipe Conference,* Tokyo, Japan, pp. 12–16.

El-Genk, M. S. & Saber, H. H., 1999. Determination of Operation Envelopes for Closed, Two-Phase Thermosyphons. *International Journal of Heat and Mass Transfer,* 42(5), 889–903. DOI: 10.1016/S0017-9310(98)00212-9.

Fadhl, B., Wrobel, L. & Jouhara, H., 2013. Numerical Modelling of the Temperature Distribution in a Two-Phase Closed Thermosyphon. *Applied Thermal Engineering,* 60, 122–131.

Faghri, A., 1986. Vapor Flow Analysis in a Double-Walled Concentric Heat Pipe. *Numerical Heat Transfer,* 10(6), 583–595.

Faghri, A., 1989. Performance Characteristics of a Concentric Annular Heat Pipe: Part II-Vapor Flow Analysis. *Journal of Heat Transfer,* 111(4), 851–857. DOI: 10.1115/1.3250796.

Faghri, A., 1995. *Heat Pipe Science and Technology.* 1st ed. Washington, DC: Taylor & Francis.

Faghri, A., 2012. Review and Advances in Heat Pipe Science and Technology. *Journal of Heat Transfer,* 134(12), 123001. DOI: 10.1115/1.4007407.

Faghri, A., 2014. Heat Pipes: Review, Opportunities and Challenges. *Frontiers in Heat Pipes (FHP)*, 5(1). DOI: 10.5098/fhp.5.1.

Faghri, A. & Buchko, M., 1991. Experimental and Numerical Analysis of Low-Temperature Heat Pipes with Multiple Heat Sources. *Journal of Heat Transfer*, 1113(3), 728–734. DOI: 10.1115/1.2910624.

Faghri, A., Buchko, M. & Cao, Y., 1991a. A Study of High-Temperature Heat Pipes with Multiple Heat Sources and Sinks: Part I—Experimental Methodology and Frozen Startup Profiles. *Journal of Heat Transfer*, 113(4), 1003–1009. DOI: 10.1115/1.2911193.

Faghri, A., Buchko, M. & Cao, Y., 1991b. A Study of High-Temperature Heat Pipes with Multiple Heat Sources and Sinks: Part II—Analysis of Continuum Transient and Steady—State Experimental Data with Numerical Predictions. *Transactions of the ASME Journal of Heat Transfer*, 113(4), 1010–1016. DOI: 10.1115/1.2911194.

Faghri, A., Chen, M. M. & Morgan, M., 1989. Heat Transfer Characteristics in Two-Phase Closed Conventional and Concentric Annular Thermosyphons. *Journal of Heat Transfer*, 111(3), 611–618. DOI: 10.1115/1.3250726.

Faghri, A., Gogineni, S. & Thomas, S., 1993. Vapor Flow Analysis of an Axially Rotating Heat Pipe. *International Journal of Heat and Mass Transfer*, 36(9), 2293–2303. DOI: 10.1016/S0017-9310(05)80114-0.

Faghri, A. & Khrustalev, D., 1997. Advances in Modeling of Enhanced Flat Miniature Heat Pipes with Capillary Grooves. *Journal of Enhanced Heat Transfer*, 4(2), 99–109.

Faghri, A. & Parvani, S., 1988. Numerical Analysis of Laminar Flow in a Double-Walled Annular Heat Pipe. *Journal of Thermophysics and Heat Transfer*, 2(3), 165–171. DOI: 10.2514/3.81.

Faghri, A. & Zhang, Y., 2006. *Transport Phenomena in Multiphase Systems*. Elsevier, ISBN-13: 978-0-12-370610-2.

Furukawa, M., 2006. Model-Based Method of Theoretical Design Analysis of a Loop Heat Pipe. *Journal of Thermophysics and Heat Transfer*, 20(1), 111–121. DOI: 10.2514/1.14675.

Ghanbarpour, M. & Khodabandeh, R., 2015. Entropy Generation Analysis of Cylindrical Heat Pipe Using Nanofluid. *Thermochimica Acta*, 610, 37–46. DOI: 10.1016/j.tca.2015.04.028.

Gray, V. H., 1969. The Rotating Heat Pipe—A Wickless, Hollow Shaft for Transferring High Heat Fluxes. *Proceedings of ASME/AIChE Heat Transfer Conference*, Minneapolis, MN, pp. 1–5.

Hall, M. L., Merrigan, M. A. & Reid, R. S., 1994. Status Report on the THROHPUT Transient Heat Pipe Modeling Code. AIP-CP-301, *Proceedings of the 11th Symposium on Space Nuclear Power and Propulsion*, American Institute of Physics, New York, NY, pp. 965–970.

Hamdan, M. O., 2003. Loop Heat Pipe (LHP) Modeling and Development by Utilizing Coherent Porous Silicon (CPS) Wicks. Ph.D. Thesis, University of Cincinnati, Cincinnati, OH.

Harley, C. & Faghri, A., 1994a. Transient Gas-Loaded Thermosyphon Analysis. *Proceedings of the 10th International Heat Transfer Conference*, Brighton, England.

Harley, C. & Faghri, A., 1994b. Complete Transient Two-Dimensional Analysis of Two-Phase Closed Thermosyphons Including the Falling Condensate Film. *Journal of Heat Transfer*, 116(2), 418–426. DOI: 10.1115/1.2911414.

Harley, C. & Faghri, A., 1995. Two-Dimensional Rotating Heat Pipe Analysis. *Journal of Heat Transfer*, 117(1), 202–208. DOI: 10.1115/1.2822304.

Harley, C. & Faghri, A., 2000. Transient Gas-Loaded Rotating Heat Pipes. *Proceedings of the 15th National and 4th ISHMT/ASME Heat and Mass Transfer Conference*, Pune, India.

Hoang, T. & Ku, J., 2003. *Transient Modeling of Loop Heat Pipes*. AIAA Paper No. 2003–6082.

Hoang, T. T. & Kaya, T., 1999. *Mathematical Modeling of Loop Heat Pipe with Two-Phase Pressure Drop*. AIAA Paper No. 1999-3448.

Holley, B. & Faghri, A., 2005. Analysis of Pulsating Heat Pipe with Capillary Wick and Varying Channel Diameter. *International Journal of Heat and Mass Transfer*, 48(13), 2635–2651. DOI: 10.1016/j.ijheatmasstransfer.2005.01.013.

Hopkins, R., Faghri, A. & Khrustalev, D., 1999. Flat Miniature Heat Pipes with Micro Capillary Grooves. *Journal of Heat Transfer*, 121(1), 102–109. DOI: 10.1115/1.2825922.

Hosoda, M., Nishio, S. & Shirakashi, R., 1999. Meandering Closed Loop Heat Transport Tube (Propagation Phenomena of Vapor Plug). AJTE99-6306, *Proceedings of the 5th ASME/JSME Joint Thermal Engineering Conference*, San Diego, CA, pp. 1–6.

Ivanovskii, M. N., Sorokin, V. P. & Yagodkin, I. V., 1982. *The Physical Principles of Heat Pipes*. Oxford: Clarendon Press.

Jang, J. H., Faghri, A., Chang, W. S. & Mahefkey, E. T., 1990. Mathematical Modeling and Analysis of Heat Pipe Start-Up from the Frozen State. *Journal of Heat Transfer*, 112(3), 586–594. DOI: 10.1115/1.2910427.

Jiao, B., Qiu, L. M., Gan, Z. H. & Zhang, X. B., 2012. Determination of the Operation Range of a Vertical Two-Phase Closed Thermosyphon. *Heat and Mass Transfer*, 48(6), 1043–1055. DOI: 10.1007/s00231-011-0954-x.

Kaya, T. & Goldak, J., 2006. Numerical Analysis of Heat and Mass Transfer in the Capillary Structure of a Loop Heat Pipe. *International Journal of Heat and Mass Transfer*, 49(17–18), 3211–3220. DOI: 10.1016/j.ijheatmasstransfer.2006.01.028.

Kaya, T. & Hoang, T. T., 1999. Mathematical Modeling of Loop Heat Pipes and Experimental Validation. *Journal of Thermophysics and Heat Transfer*, 13(3), 314–320. DOI: 10.2514/2.6461.

Kaya, T. & Ku, J., 1999. *A Parametric Study of Performance Characteristics of Loop Heat Pipes*. SAE Paper No. 2001-01-2317.

Kaya, T. & Ku, J., 2003. Thermal Operational Characteristics of a Small-Loop Heat Pipe. *Journal of Thermophysics and Heat Transfer*, 17(4), 464–470. DOI: 10.2514/2.6805.

Kaya, T., Pérez, R., Gregori, C. & Torres, A., 2008. Numerical Simulation of Transient Operation of Loop Heat Pipes. *Applied Thermal Engineering*, 28(8–9), 967–974. DOI: 10.1016/j.applthermaleng.2007.06.037.

Khandekar, S. & Gupta, A., 2007. Embedded Pulsating Heat Pipe Radiators. *Proceedings of the 14th International Heat Pipe Conference*, Florianopolis, Brazil, pp. 22–27.

Khandekar, S., Panigrahi, P. K., Lefèvre, F. & Bonjour, J., 2010. Local Hydrodynamics of Flow in a Pulsating Heat Pipe: A Review. *Frontiers in Heat Pipes (FHP)*, 1, 55–67, 023003. DOI: 10.5098/fhp.v1.2.3002.

Khandekar, S., Schneider, M., Schäfer, P., Kulenovic, R., and Groll, M., 2002. Thermofluid Dynamic Study of Flat-Plate Closed-Loop Pulsating Heat Pipes. *Microscale Thermophysical Engineering*, 6(4), 303–317. DOI: 10.1080/10893950290098340.

Khrustalev, D., 2010. Advances in Transient Modeling of Loop Heat Pipe Systems with Multiple Components. *AIP Conference Proceedings*, 1208, 55–67.

Khrustalev, D. & Faghri, A., 1994. Thermal Analysis of a Micro Heat Pipe. *Journal of Heat Transfer*, 116(1), 189–198. DOI: 10.1115/1.2910855.

Khrustalev, D. & Faghri, A., 1995a. Thermal Characteristics of Conventional and Flat Miniature Axially-Grooved Heat Pipes. *Journal of Heat Transfer*, 117(4), 1048–1054. DOI: 10.1115/1.2836280.

Khrustalev, D. & Faghri, A., 1995b. Heat Transfer During Evaporation on Capillary-Grooved Structures of Heat Pipes. *Journal of Heat Transfer*, 117(3), 740–747. DOI: 10.1115/1.2822638.

Kiper, A. M., Feric, G., Anjum, M. & Swanson, T. D., 1990. Transient Analysis of a Capillary Pumped Loop Heat Pipe. AIAA-90-1685, *Proceedings of AIAA/ASME 5th Joint Thermophysics and Heat Transfer Conference*, Seattle, WA.

Kroliczek, E. J., Ku, J. & Ollendorf, S., 1984. Design, Development, and Test of a Capillary Pump Heat Pipe. AIAA-84-1720, *Proceedings of AIAA 19th Thermophysics Conference,* Snowmass, CO.

Ku, J., 1993. Capillary Pump Loop for the Systems of Thermal Regulation of Spacecraft. *Proceedings of ASME National Heat Transfer Conference,* Atlanta, GA.

Ku, J., Kroliczek, E. J., Butler, D., Schweickart, R. B. and McIntosh, R., 1986a. Capillary Pumped Loop GAS and Hitchhiker Flight Experiments. AIAA-86-1249, *Proceedings of AIAA/ASME 4th Joint Thermophysics and Heat Transfer Conference,* Boston, MA.

Ku, J., Kroliczek, E. J., McCabe, M. E. & Benner, S. M., 1988. A High Power Spacecraft Thermal Management System. AIAA-88-2702, *Proceedings of AIAA Thermophysics, Plasmadynamics, and Lasers Conference,* San Antonio, TX.

Ku, J., Kroliczek, E. J. & McIntosh, R., 1987a. Analytical Modelling of the Capillary Pumped Loop. *Proceedings of the 6th International Heat Pipe Conference,* Grenoble, France.

Ku, J., Kroliczek, E. J. & McIntosh, R., 1987b. Capillary Pumped Loop Technology Development. *Proceedings of the 6th International Heat Pipe Conference,* Grenoble, France.

Ku, J., Kroliczek, E. J., Taylor, W. J. & McIntosh, R., 1986b. Functional and Performance Tests of Two Capillary Pumped Loop Engineering Models. AIAA-86-1248, *Proceedings of AIAA/ASME 4th Joint Thermophysics and Heat Transfer Conference,* Boston, MA.

Launay, S., Platel, V., Dutour, S. & Joly, J. L., 2007b. Transient Modeling of Loop Heat Pipes for the Oscillating Behavior Study. *Journal of Thermophysics and Heat Transfer,* 21(3), 487–495. DOI: 10.2514/1.26854.

Launay, S., Sartre, V. & Bonjour, J., 2007a. Effects of Fluid Thermophysical Properties on Loop Heat Pipe Operation. *Proceedings of the 14th International Heat Pipe Conference,* Florianopolis, Brazil.

Launay, S., Sartre, V. & Bonjour, J., 2007c. Parametric Analysis of Loop Heat Pipe Operation: A Literature Review. *International Journal of Thermal Sciences,* 46(7), 621–636. DOI: 10.1016/j.ijthermalsci.2006.11.007.

Launay, S., Sartre, V. & Bonjour, J., 2008. Analytical Model for Characterization of Loop Heat Pipes. *Journal of Thermophysics and Heat Transfer,* 22(4), 623–631. DOI: 10.2514/1.37439.

Launay, S., Sartre, V., Mantelli, M. H., De Paiva, K. V. & Lallemand, M., 2004. Investigation of a Wire Plate Micro Heat Pipe Array. *International Journal of Thermal Sciences,* 43(5), 499–507. DOI: 10.1016/j.ijthermalsci.2003.10.006.

Lefèvre, F. & Lallemand, M., 2006. Coupled Thermal and Hydrodynamic Models of Flat Micro Heat Pipes for the Cooling of Multiple Electronic Components. *International Journal of Heat and Mass Transfer,* 49(7–8), 1375–1383. DOI: 10.1016/j.ijheatmasstransfer.2005.10.001.

Lefèvre, F., Lips, S. & Bonjour, J., 2010. Investigation of Evaporation and Condensation Processes Specific to Grooved Flat Heat Pipes. *Frontiers in Heat Pipes (FHP),* 1, 023001. DOI: 10.5098/fhp.v1.2.3001.

Liang, S. B. & Ma, H. B., 2004. Oscillating Motions of Slug Flow in Capillary Tubes. *International Communications in Heat and Mass Transfer,* 31(3), 365–375. DOI: 10.1016/j.icheatmasstransfer.2004.02.007.

Lin, L. & Faghri, A., 1997a. Heat Transfer Analysis of Stratified Flow in Rotating Heat Pipes with Cylindrical and Stepped Walls. *International Journal of Heat and Mass Transfer,* 40(18), 4393–4404. DOI: 10.1016/S0017-9310(97)00060-4.

Lin, L. & Faghri, A., 1997b. Steady-State Performance in a Thermosyphon with Tube Separator. *Applied Thermal Engineering,* 667–679. DOI: 10.1016/S1359-4311(96)00084-1.

Lin, L. & Faghri, A., 1997c. Steady-State Performance of a Rotating Miniature Heat Pipe. *Journal of Thermophysics and Heat Transfer,* 11(4), 513–518. DOI: 10.2514/2.6292.

Lin, L. & Faghri, A., 1998a. Condensation in Rotating Stepped Wall Heat Pipes with Hysteretic Annular Flow. *Journal of Thermophysics and Heat Transfer,* 12(1), 94–99. DOI: 10.2514/2.6307.

Lin, L. & Faghri, A., 1998b. An Analysis of Two-Phase Flow Stability in a Thermosyphon with Tube Separator. *Applied Thermal Engineering,* 18(6), 441–455. DOI: 10.1016/S1359-4311(97)00046-X.

Lin, L. & Faghri, A., 1999. Heat Transfer in Micro Region of a Rotating Miniature Heat Pipe. *International Journal of Heat and Mass Transfer,* 42(8), 1363–1369. DOI: 10.1016/S0017-9310(98)00270-1.

Ma, H. B., Hanlon, M. A. & Chen, C. L., 2006. An Investigation of Oscillating Motions in a Miniature Pulsating Heat Pip. *Microfluidics and Nanofluidics,* 2(2), 171–179. DOI: 10.1007/s10404-005-0061-8.

Ma, H. B., Maschmann, M. R. & Liang, S. B., 2002. *Heat Transport Capability in a Pulsating Heat Pipe.* AIAA 2002-2765.

Mashaei, P. & Shahryari, M., 2015. Effect of Nanofluid on Thermal Performance of Heat Pipe with Two Evaporators; Application to Satellite Equipment Cooling. *Acta Astronautica,* 111, 345–355.

Maydanik, Y. F., Fershtater, Y. G. & Solodovnik, N., 1994. *Loop Heat Pipes: Design, Investigation, Prospects of Use in Aerospace Technics.* SAE Paper 941185.

Miyazaki, Y. & Akachi, H., 1996. Heat Transfer Characteristics of Looped Capillary Heat Pipe. *Proceedings of the 5th International Heat Pipe Symposium,* Melbourne, Australia, pp. 378–383.

Miyazaki, Y. & Akachi, H., 1998. Self Excited Oscillation of Slug Flow in a Micro Channel. *Proceedings of the 3rd International Conference on Multiphase Flow,* Lyon, France.

Miyazaki, Y. & Arikawa, M., 1999. Oscillatory Flow in the Oscillating Heat Pipe. *Proceedings of the 11th International Heat Pipe Conference,* Tokyo, Japan, pp. 131–136.

Muraoka, I., Ramos, F. M. & Vlassov, V. V., 2001. Analysis of the Operational Characteristics and Limits of a Loop Heat Pipe with Porous Element in the Condenser. *International Journal of Heat and Mass Transfer,* 44(12), 2287–2297. DOI: 10.1016/S0017-9310(00)00259-3.

Nithyanandam, K. & Pitchumani, R., 2013. Thermal Energy Storage with Heat Transfer Augmentation using Thermosyphons. *International Journal of Heat Mass Transfer,* 67, 281–294.

Pan, Y., 2001. Condensation Heat Transfer Characteristics and Concept of Sub-Flooding Limit in a Two-Phase Closed Thermosyphon. *International Communications in Heat and Mass Transfer,* 28(3), 311–322. DOI: 10.1016/S0735-1933(01)00237-8.

Ponnappan, R., 1990. Comparison of Vacuum and Gas Loaded Mode Performances of a LMHP. AIAA-90-1755, *Proceedings of AIAA/ASME 5th Joint Thermophysics and Heat Transfer Conference,* Seattle, WA.

Ranjan, R., Murthy, J. Y., Garimella, S. V. & Vadakkan, U., 2011. A Numerical Model for Transport in Flat Heat Pipes Considering Wick Microstructure Effects. *International Journal of Heat and Mass Transfer,* 54(1–3), 153–168. DOI: 10.1016/j.ijheatmasstransfer.2010.09.057.

Rice, J. & Faghri, A., 2007. Analysis of Porous Wick Heat Pipes, Including Capillary Dry-Out Limitations. *AIAA Journal of Thermophysics and Heat Transfer,* 21(3), 475–486. DOI: 10.2514/1.24809.

Rosenfeld, J. H., 1987. Modeling of Heat Transfer into a Heat Pipe for a Localized Heat Input Zone. *AIChE Symposium Series,* 83, 71–76.

Sakulchangsatjatai, P., Terdtoon, P., Wongratanaphisan, T., Kamonpet, P. & Murakami, M., 2004. Operation Modeling of Closed-End and Closed-Loop Oscillating Heat Pipes at Normal Operating Condition. *Applied Thermal Engineering,* 24(7), 995–1008. DOI: 10.1016/j.applthermaleng.2003.11.006.

Sartre, V., Zaghdoudi, M. C. & Lallemand, M., 2000. Effect of Interfacial Phenomena on Evaporative Heat Transfer in Micro Heat Pipes. *International Journal of Thermal Sciences*, 39(4), 498–504. DOI: 10.1016/S1290-0729(00)00205-2.

Schweickart, R. B. & Buchko, M. T., 1991. Development and Test Results of a Two-Phase Reservoir for Thermal Transport Systems used in Micro-Gravity. *SAE (Society of Automotive Engineers) Transactions*, 100, 1887–1899.

Shabgard, H., Allen, M. J., Sharifi, N., Benn, S. P., Faghri, A. & Bergman, T. L., 2015. Heat Pipe Heat Exchangers and Heat Sinks: Opportunities, Challenges, Applications, Analysis, and State of the Art. *International Journal of Heat and Mass Transfer*, 89, 138–158. DOI: 10.1016/j.ijheatmasstransfer.2015.05.020.

Shabgard, H., Bergman, T., Sharifi, N. & Faghri, A., 2010. High Temperature Latent Heat Thermal Energy Storage Using Heat Pipes. *International Journal of Heat Mass Transfer*, 53, 2979–2988. DOI: 10.1016/j.ijheatmasstransfer.2010.03.035.

Shabgard, H. & Faghri, A., 2011. Performance Characteristics of Cylindrical Heat Pipes with Multiple Heat Sources. *Applied Thermal Engineering*, 31(16), 3410–3419. DOI: 10.1016/j.applthermaleng.2011.06.026.

Shabgard, H., Xiao, B., Faghri, A., Gupta, R. & Weissman, W., 2014. Thermal Characteristics of a Closed Thermosyphon Under Various Filling Conditions. *International Journal of Heat and Mass Transfer*, 70, 91–102. DOI: 10.1016/j.ijheatmasstransfer.2013.10.053.

Shafahi, M., Bianco, V., Vafai, K. & Manca, O., 2010. An Investigation of the Thermal Performance of Cylindrical Heat Pipes Using Nanofluids. *International Journal of Heat and Mass Transfer*, 53, 376–383. DOI: 10.1016/j.ijheatmasstransfer.2009.09.019.

Shafii, M. B., Faghri, A. & Zhang, Y., 2001. Thermal Modeling of Unlooped and Looped Pulsating Heat Pipes. *Journal of Heat Transfer*, 123(6), 1159–1172. DOI: 10.1115/1.1409266.

Shafii, M. B., Faghri, A. & Zhang, Y., 2002. Analysis of Heat Transfer in Unlooped and Looped Pulsating Heat Pipes. *International Journal of Numerical Methods for Heat and Fluid Flow*, 12(5), 585–609. DOI: 10.1108/09615530210434304.

Sharifi, N., Wang, S., Bergman, T. & Faghri, A., 2012. Heat Pipe-Assisted Melting of a Phase Change Material. *International Journal of Heat Mass Transfer*, 55, 3458–3469. DOI: 10.1016/j.ijheatmasstransfer.2012.03.023.

Spendel, T., 1984. Laminar Film Condensation Heat Transfer in Closed Two-Phase Thermosyphons. *Proceedings of 5th International Heat Pipe Conference*, Tsukuba, Japan.

Suman, B. & Kumar, P., 2005. An Analytical Model for Fluid Flow and Heat Transfer in a Micro-Heat Pipe of Polygonal Shape. *International Journal of Heat and Mass Transfer*, 48(21–22), 4498–4509. DOI: 10.1016/j.ijheatmasstransfer.2005.05.001.

Tournier, J. M. & El-Genk, M. S., 1994. A Heat Pipe Transient Analysis Model. *International Journal of Heat and Mass Transfer*, 37(5), 753–762. DOI: 10.1016/0017-9310(94)90113-9.

Tournier, J. M. & El-Genk, M. S., 1996. Vapor Flow Model for Analysis of Liquid-Metal Heat Pipe Startup from a Frozen State. *International Journal of Heat and Mass Transfer*, 39(18), 3767–3780. DOI: 10.1016/0017-9310(96)00066-X.

Wang, Y. & Vafai, K., 2000. An Experimental Investigation of the Transient Characteristics on a Flat-Plate Heat Pipe during Startup and Shutdown Operations. *Journal of Heat Transfer*, 122(3), 525–535. DOI: 10.1115/1.1287725.

Wang, Y. X. & Peterson, G. P., 2002. Analysis of Wire-Bonded Micro Heat Pipe Arrays. *Journal of Thermophysics and Heat Transfer*, 16(3), 346–355. DOI: 10.2514/2.6711.

Wong, T. N., Tong, B. Y., Lim, S. M. & Ooi, K. T., 1999. Theoretical Modeling of Pulsating Heat Pipe. *Proceedings of the 11th International Heat Pipe Conference*, Tokyo, Japan, pp. 159–163.

Xiao, B. & Faghri, A., 2008. A Three-Dimensional Thermal-Fluid Analysis of Flat Heat Pipes. *International Journal of Heat and Mass Transfer*, 51(11–12), 3113–3126. DOI: 10.1016/j.ijheatmasstransfer.2007.08.023.

Zhang, Y. & Faghri, A., 2002. Heat Transfer in a Pulsating Heat Pipe with Open End. *International Journal of Heat and Mass Transfer*, 45(4), 755–764. DOI: 10.1016/ S0017-9310(01)00203-4.

Zhang, Y. & Faghri, A., 2003. Oscillatory Flow in Pulsating Heat Pipes with Arbitrary Numbers of Turns. *Journal of Thermophysics and Heat Transfer*, 17(3), 340–347. DOI: 10.2514/2.6791.

Zhang, Y. & Faghri, A., 2008. Advances and Unsolved Issues in Pulsating Heat Pipes. *Heat Transfer Engineering*, 29(1), 20–44. DOI: 10.1080/01457630701677114.

Zhang, Y., Faghri, A. & Shafii, M. B., 2002. Analysis of Liquid-Vapor Pulsating Flow in a U-Shaped Miniature Tube. *International Journal of Heat and Mass Transfer*, 45(12), 2501–2508. DOI: 10.1016/S0017-9310(01)00348-9.

Zhu, N. & Vafai, K., 1998a. Analytical Modeling of the Startup Characteristics of Asymmetrical Flat-Plate and Disk-Shaped Heat Pipes. *International Journal of Heat and Mass Transfer*, 41(17), 2619–2637. DOI: 10.1016/S0017-9310(97)00325-6.

Zhu, N. & Vafai, K., 1998b. Vapor and Liquid Flow in an Asymmetrical Flat Plate Heat Pipe: A Three-Dimensional Analytical and Numerical Investigation. *International Journal of Heat and Mass Transfer*, 41(1), 159–174. DOI: 10.1016/S0017-9310(97)00075-6.

Zuo, Z.J. & Faghri, A., 1998. A Network Thermodynamic Analysis of the Heat Pipe. *International Journal of Heat and Mass Transfer*, 41(11), 1473–1484. DOI: 10.1016/ S0017-9310(97)00220-2.

Zuo, Z. J. & Gunnerson, F. S., 1995. Heat Transfer Analysis of an Inclined Two-Phase Closed Thermosyphon. *Journal of Heat Transfer*, 117(4), 1073–1075. DOI: 10.1115/1.2836287.

Zuo, Z. J., North, M. T. & Ray, L., 1999. Combined Pulsating and Capillary Heat Pipe Mechanism for Cooling of High Heat Flux Electronics. *Proceedings of ASME Heat Transfer Device Conference*, Nashville, TN, pp. 2237–2243.

Zuo, Z. J., North, M. T. & Wert, K. L., 2001. High Heat Flux Heat Pipe Mechanism for Cooling of Electronics. *IEEE Transactions on Components and Packaging Technologies*, 24(2), 220–225. DOI: 10.1109/6144.926386.

Index

Printed in the United States
by Baker & Taylor Publisher Services